THE HERSCHEL PARTNERSHIP

WILLIAM HERSCHEL (1738–1822), a talented musician like his father and brothers, was born in Hanover, but fled to England in 1757 as a refugee from the Seven Years War. In 1766 he accepted a post as organist in Bath, the fashionable spa resort in the west of England where there were rich pickings for performers and teachers of music. Six years later he rescued his sister CAROLINE (1750–1848) from servitude in the family home, and took her to Bath where she was to run his household and see whether she could make a career as a singer. Soon however William became a keen amateur astronomer, and in 1781 he achieved fame as the discoverer of the planet Uranus. Next year the King conferred on him a pension that would enable him to give up music and devote himself to astronomy, his only duties being to reside near Windsor Castle so as to be available to show the heavens to the Royal Family and their guests when so requested. This required Caroline to abandon her career as a singer and become his assistant. They then embarked on a partnership unequalled in the history of astronomy: William observed, while Caroline sat at a desk nearby writing down the observations as her brother shouted them out. In the morning she would write up a fair copy of the night's work, and in due course she would prepare William's papers for publication. She also observed on her own account, and earned fame as the discoverer of eight comets. But in 1788 William married and Caroline found herself displaced in his affections. She was however compensated with a salary from the Crown for her work as his assistant, so becoming the first female in history to earn her living from astronomy. This is the story of their partnership, as viewed through the eyes of Caroline.

MICHAEL HOSKIN before retirement was Head of the Department of History and Philosophy of Science at Cambridge University. In 1970 he founded *Journal for the History of Astronomy*, which he has edited ever since. In 1982 he gave an Invited Discourse to the International Astronomical Union in the Roman Odeon at Patras, Greece, and in 2001 the Union named a minor planet in his honour. He wrote his first book on the Herschels in 1959.

"Her fame will be held in honour throughout all ages" — Pierre-François-André Méchain, 1789. The only known portrait of Caroline Herschel as a young woman, painted before she left Hanover in 1772 to join William in Bath. Courtesy of the Museum of the History of Science, Oxford University.

THE HERSCHEL PARTNERSHIP

as viewed by Caroline

Michael Hoskin

SCIENCE HISTORY PUBLICATIONS

Published by Science History Publications Ltd
16 Rutherford Road, Cambridge, CB2 2HH
United Kingdom
www.shpltd.co.uk

Copyright © 2003 Science History Publications Ltd

First published 2003

Printed and bound in the United Kingdom by
University Printing Services, University Press, Cambridge

ISBN 0 905193 05 9

Contents

Preface

For this study I am more than usually indebted to friends old and new. My colleague Anthony J. Turner at one time planned an edition of Caroline's autobiographies, and assembled materials for the purpose; in a signal act of generosity he presented these to me on indefinite loan, and this book has been greatly enriched thereby. I am grateful also to Richard Phillips, Michael Tabb and Trevor Fawcett, who shared their knowledge of the Herschel homes in Bath, and to Janet Kennish, a local historian of Datchet, and Anthony Fanning of Old Windsor, who did the same for the Herschel homes in the vicinity of Windsor. Arndt Latusseck of Hanover investigated archives in Germany, established the chronology of the French occupation of Hanover, and found testimony to the murder by strangulation of Jacob Herschel. Kenneth Edward James, historian of musical life in Bath, shared his knowledge and much of his thesis with me. My colleague Bradley E. Schaefer found time to examine for me Herschel manuscripts now at the University of Texas at Austin, while Patricia Fara, Owen Gingerich and Anthony J. Turner kindly commented on an early draft of this book.

The William Herschel Museum, which is located in one of the houses at Bath associated with William and Caroline, owns the original typescript of *The Herschel Chronicle* by William's granddaughter, Constance A. Lubbock. The *Chronicle* is required reading for students of the Herschels, but in fact Lady Lubbock submitted to Cambridge University Press an altogether longer text, which the press editor dutifully condensed and rationalized for publication. The result is a better book but an emasculated source for historians. The Museum (which is the home of the William Herschel Society) generously allowed me to have the typescript on extended loan.

I am grateful to the Librarians of the Royal Society, the Royal Astronomical Society, and the Harry Ransom Library of the Humanities Research Center of the University of Austin at Texas, for access to materials in their care; and I am particularly grateful to Peter D. Hingley and the Royal Astronomical Society for permission to reproduce many of the illustrations in this book.

All students of the Herschels seek help sooner or later from William's descendent, John Herschel-Shorland, and never without success. His kindness in the cause of Herschel scholarship knows no bounds.

Lastly, I thank Bernard Hoskin, who used his skills in electronic publishing to convert my typescript into this book.

CHILDREN BORN TO ISAAC AND ANNA HERSCHEL

Sophia Elisabeth [Sophia]
born 12 April 1733, died April 1803

Heinrich Anton Jacob [Jacob]
born 20 November 1734, murdered 1792

Johann Heinrich
born 25 April 1736, died 26 October 1743

Friedrich Wilhelm [William]
born 15 November 1738, died 25 August 1822

Anna Christina
born 12/13 July 1741, died (of whooping cough) 22 July 1748

Johann Alexander [Alexander]
born 13 November 1745, died 15 March 1821

Maria Dorethea
born 8 June 1748, died 21 April 1749

Carolina Lucretia [Caroline]
born 16 March 1750, died 9 January 1848

Frantz Johann
born 5 February [or 13 May] 1752,
died (of smallpox) 26 March [or 6 May] 1754

Johann Dietrich [Dietrich]
born 13 September 1755, died late January 1827

Introduction

In 1787, King George III agreed to pay £50 a year to Caroline Herschel for her work as assistant to his astronomer, her brother William, and Caroline thereby became the first salaried female in the history of astronomy. In an age when women were excluded from science, she played a crucial role in a partnership that altered the course of its history. This is the story of that partnership, as seen through the eyes of Caroline.

In the century before Caroline's birth in 1750, at least four women had succeeded in making modest contributions to astronomy; and, like Caroline, all relied on family connections to overcome the obstacles due to their sex. The most obscure and the most remarkable was Maria Cunitz (1610–64), the daughter of a wealthy Silesian landowner who had her taught medicine and mathematics as well as six languages. Her family took refuge in Poland during the Thirty Years War, and it was there that she prepared astronomical tables based on Kepler's Rudolphine Tables of 1627. Her folio volume *Urania Propitia sive Tabulae Astronomicae mirè faciles, vim hypothesium physicarum a Kepplero proditarum complexae*, was privately published in 1650.

The best-known of the four is Catherina Elisabetha Koopman (1647–93), second wife of the seventeenth-century Danzig brewer Johannes Hevelius.[1] As an amateur astronomer, Hevelius was primarily an observer, indeed the last great observer to refuse to incorporate telescope sights into his measuring instruments. Elisabetha acted as hostess to visiting astronomers such as Edmond Halley, she edited some of her husband's works that were still unpublished when he died in 1687, and she is shown in engravings as a partner in his observations.

The remaining two were respectively the wife and daughter of Gottfried Kirch, head of the Berlin Observatory founded in 1700 by the future Frederick I of Prussia.[2] Maria Margarethe Kirch (1670–1720) helped her husband with observations, and herself discovered the comet of 1702. She had a talent for the calculation of calendars, which she shared with her daughter Christine (*c.* 1696–1782). After Gottfried's death in 1710, Maria helped in other observatories, and was invited to Russia by Peter the Great; but when her son Christfried became observer at Berlin in 1716, she joined him there. Both she and Christine assisted Christfried with his

Johannes Hevelius observing with the help of his wife, Elisabetha. From Hevelius's *Machinae coelestis pars prior* (Danzig, 1673).

observations and calculations.

These women are remembered because it was then almost unknown for a female to contribute to science. The same was true a century later when William Herschel, a musician and amateur astronomer based at Bath in the west of England, burst upon the scene with his discovery in 1781 of the planet Uranus. Within months his achievement had brought him royal patronage, and this would allow him to give up music and devote himself to astronomy. William was unmarried, and Caroline had been running his home, as well as singing in his concerts. There were to be no more concerts. Instead, brother and sister moved to be within reach of Windsor Castle, where, against all convention, Caroline "found that I was to be trained for an assistant Astronomer".

Contemporary astronomers, both professional and amateur, were preoccupied with the solar system — the Sun, planets and their satellites, and comets. The distant stars changed little from one century to the next; to observers they were a convenient backcloth to movements within the solar system, reference points for the positions of the Moon and planets, and whole books on astronomy were written with barely a mention of stars or indeed of the universe as a whole. It was what happened within a few light-minutes of Earth that was all-important.

William, being self-taught, did not know this. From his earliest days as an amateur astronomer he set out to explore "the construction of the heavens", the large-scale structure of the universe. To see distant and therefore faint objects he would need telescopes with mirrors as big as possible — with the maximum "light-gathering power" — and he was to build himself the biggest reflecting telescopes the world had yet seen; not only that, but professional astronomers would compete with the crowned heads of Europe for the privilege of buying one of the instruments that William offered for sale.

Equipped with his cosmological artillery, William set out on some of the greatest observing campaigns of astronomical history. He spent twenty years sweeping the night sky, region by region, in his search for the milky patches known as nebulae. A hundred or so were known when he started, two-and-a-half thousand when he finished. This embarrassment of riches itself made a reorientation of astronomy almost inevitable. The members of the solar system were few in number, familiar individuals with personal names, orbiting endlessly round as though attached to the cogs of a vast clock; indeed, God was the great clockmaker. But there was no question of giving personal names to each of thousands of nebulae: William had inevitably to become a natural historian of the heavens, collecting specimens in great numbers and classifying them, an activity hitherto unknown in astronomy.

But how to classify them? Many nebulae proved under scrutiny to be nothing more than clusters of thousands of stars. If stars were clustered together in this way, there had to be some force at work to bring this about; they must be in each other's grasp, so to speak, attracting each other by Newtonian gravity or some similar force. But if so, then — as time passed — a once-scattered cluster would become more and more compressed.

In place of a clockwork universe, therefore, William substituted one whose components underwent changes over time analogous to the life-cycles of living creatures. He classified a scattered cluster as young, a more compressed one as middle-aged, a tightly compressed one as nearing the end of its life cycle. He redirected astronomy towards the world-picture that astronomers know today, in which there is development over time for everything from an individual star to the universe as a whole. William not only built the greatest telescopes of his day, and used them in observational campaigns the equal of any in the history of astronomy, but he deployed the data that resulted to reorientate the entire science.

Caroline's assistance in the twenty-year search for nebulae was indispensable because William found it impossible to both observe and record. Before each observing session he had to wait until his eyesight had become fully adjusted to the dark, for most nebulae are very faint; if, having found a nebula, he then went into the light and wrote down what he had seen, he would have to wait minutes more before his eyes were again dark-adjusted and he could continue his observations. The solution lay in a partnership: William would do the observing, and Caroline would do the recording. William at the eyepiece of the telescope would shout out a description of what he could see; Caroline, seated at a desk at an open window nearby, would write down his description, together with the time and the angle at which the telescope tube was elevated, and then call back the description as confirmation that she had heard her brother correctly.

This was only the first of Caroline's many contributions to their partnership. Come the morning, she would make a fair copy of the night's observations, and in due course she would carry out the routine calculations and assemble the data for publication. If William was lent an important paper to study, Caroline would transcribe it for him. When William prepared for publication one of his many theoretical papers, written down on scraps of paper, Caroline would write up one fair copy for their files and another for *Philosophical Transactions*. Whatever her brother needed her to do, she gladly did. No hired hand, she was at his disposal around the clock, ready to help in any way she could.

As children, William and Caroline had been taught the Protestant work ethic: "it was only in the thought of having done our duty, that we could expect to find contentment."[3] Despite her multifarious duties, Caroline still had uncommitted night hours, and so her brother assigned her to 'sweep' the sky for interesting objects, and especially for newly arriving comets; and he made her telescopes designed for this express purpose. Over the years Caroline found no fewer than eight comets, and she established herself as one of the great comet-hunters of her day. Male astronomers worldwide were astonished, and ransacked their vocabularies for adequate compliments.

That Caroline played a crucial role in a partnership that marks an epoch in the history of astronomy is beyond question. But was she herself an assistant astronomer, or merely a devoted astronomer's assistant? Perhaps the pages that follow will provide the answer.

Hanoverian Prologue

It is the autumn of 1757. The Seven Years War is casting a long shadow over Europe, and the town of Hanover is in panic. The army of the Elector of Hanover, better known by his other title of King George II of Great Britain, has been defeated by the French at nearby Hastenbeck, its troops are "flying or scattered in all directions",[1] and the French may appear at the gates of the city at any time.

In one of the poorer homes a seven-year-old girl called Caroline is sitting on the front doorstep, wondering why she has constantly been kept out of the way. Then she glimpses the passing figure of a young man, shrouded in a greatcoat and making himself as inconspicuous as possible. She recognizes the greatcoat as belonging to their landlord,[2] but the wearer is her favourite brother, William, who has been a bandsman of the Hanoverian Guards since he entered four years ago at the age of fourteen.

William is carrying little, apparently a man innocently on his way from one part of town to another. After he disappears from sight, their mother follows in his tracks, carrying a nondescript bundle in which his uniform is hidden. As she is a woman she will not be challenged.

She returns an hour or so later, mission accomplished, "alife, but heart-broken".[3] William has succeeded in slipping past the last sentinel, and so is safe at last from being pressed into the makeshift army the burghers are raising for the defence of Hanover.[4] She tells how she followed him until they were well outside the town, gave him his uniform, and stood watching until he disappeared from sight, a son on his way to rejoin a regiment at war.[5] In less than three months he will be in England, a refugee.

William will return one day to collect his little sister, and take her to live with him in his adopted country. Together they will earn undying fame, as the greatest partnership the history of astronomy has yet known.

1

A Troubled Childhood

Marriages, it is said, are often unions of opposites, and that celebrated in Hanover on 12 October 1732[1] between Isaac Herschel and Anna Ilse Moritzen of nearby Neustadt was no exception. Isaac was a gardener who had turned himself into a professional musician by a combination of talent and determination.[2] Anna had neither ambition nor ability; as a child she had attended school for a while, but like most girl pupils in country schools,[3] when she left she could neither read nor write.[4] Isaac wanted his children to have the best education he could afford, and took pleasure in helping them with their lessons after they came home from school.[5] Anna on the other hand regarded a little learning as a dangerous thing; and a lot of learning, as still more dangerous. Looking back in later life, she would feel vindicated: with less education, her sons might have been content to remain in their native Hanover.[6]

As a boy Isaac had seemed fated to be a gardener. His father was a gardener before him, and his older brother served an apprenticeship in the same trade before turning to farming. But before Isaac was old enough to be apprenticed his father died,[7] leaving his mother too poor to make the necessary payments. Despite this, Isaac managed to learn the rudiments of gardening; and before long his brother found him a job, tending the garden of an elderly widow.[8]

But Isaac found the life of a gardener wholly unsatisfying. Although there was no tradition of music in his family, he had somehow acquired musical genes in abundance: five of his six surviving children would become professional musicians or singers, and the sixth, a daughter, had five sons all of whom did the same.[9] While still living at home with his mother, Isaac had managed to lay hands on a violin, and he taught himself to play on it by ear.[10] Later he used his wages from the widow to pay an oboist for music lessons. He bought himself an oboe, and practised "day and night as much as I could get time, as my wish was to become an oboist; for I had lost all taste for gardening".[11]

When he was 21 Isaac bade farewell to the widow and set out to try his luck as a musician. At Berlin he looked into the possibility of joining a regimental band, but on further consideration decided the service would be "very bad and slavish".[12] With help from his brother and sister he then studied for a year in Potsdam with a Prussian conductor and oboist.[13] But his cautious brother was sceptical of Isaac's prospects as a musician: attending recitals is optional, but everyone has to eat. Isaac should give up music and come and live with his brother, and learn the secure trade of a farmer.[14]

Isaac saw the force of this argument, and allowed himself to be persuaded; but it was not long before the siren call of music proved too strong. In July 1731 he made the journey to Brunswick, where he was offered a post as oboist, though the independent young man declined it as "too Prussian".[15] The following month he arrived in Hanover, and within days had found a position to his liking, as oboist in the Foot Guards based in the town.[16] A year later, on 12 October, he and Anna were married. Why a cultivated man like Isaac chose to marry the illiterate Anna is not hard to seek: she was already several weeks pregnant with their daughter Sophia.[17]

Isaac and Anna were to have nine more children, four of whom died young. As in many marriages, the father took responsibility for the upbringing of the boys and the mother of the girls. All the children attended the Garrison school until they were fourteen, but the education there was rudimentary. If William's recollection in his old age is to be trusted, there were no fewer than five hundred boys there, and both William and Caroline imply that there was but one teacher for the entire school. Certainly the ablest of the older pupils were expected to instruct the younger.[18] William is careful to say that "all the boys received lessons in reading, writing and arithmetic",[19] but as a girl Caroline was taught simply "to read and write and be informed of our religious duties".[20] She makes no mention of arithmetic; and when as a young adult she joined her brother William in England, he had to teach her the addition and subtraction necessary for keeping the household accounts. In later life she would carry the multiplication tables around with her.[21]

Isaac's ambitions for his sons far exceeded the resources of the Garrison school. Although he reached the limit of his own abilities when he became leader of the army band,[22] he was an enthusiastic and gifted teacher of music,[23] not least when the pupils were his own children.[24] As soon as the boys were old enough to hold a small violin their lessons began; all showed precocious talent[25] and the four who survived childhood went on to distinguished careers. Not only that, but Isaac was a remarkable man of genuine breadth of mind, eager to help his children with their studies and to widen their intellectual horizons.[26]

In times of peace the prospects for the Herschel family were good, despite the

humble circumstances in which they lived. The demands on an army bandsman
were modest: Isaac could supplement his pay by giving lessons, and he was often
to be found at home with his family. But in war things were very different; Isaac
would be away with the regiment, the family would have to survive on his basic
pay, and there was no one to help with boys with their school work or — more
importantly — their musical lessons. Unfortunately, it was Isaac's fate to live in
turbulent times.

In 1740 Frederick II came to the throne of Prussia and, hell-bent on qualifying
for the title of 'the Great',[27] soon plunged Europe into war. On 6 September 1741,
by which time the number of Herschel children had risen to five, Isaac had to leave
Hanover with his regiment; they encamped at Hameln, though they were back in
Hanover six weeks later.[28] On 10 September 1742 they were off again, this time
for Braband, and the following 16 June the regiment took part in the Battle of
Dettingen. Bandsmen were always permitted to take shelter when the shooting
started, but otherwise they shared the hardships of the campaigning, and Isaac
found himself spending the night after the battle in a waterlogged ditch. This
affected his health, and early in 1744, from 10 January to 4 April, he was allowed
home to convalesce.[29]

The coming winter Isaac and his comrades were granted a much-needed spell of
leave. They arrived in Hanover on 25 January 1745, and soon Anna conceived their
sixth child, Alexander (born 13 November); but on 10 April they were off again,
this time with a contingent of recruits, bound for Koblenz. At last, on 13 February
1746, the entire regiment returned to Hanover, and normal life could resume. "Got
sy Danck!", was Isaac's heartfelt comment.[30]

The experience had been traumatic. His own health had been permanently
undermined, he had been absent from the deathbed of his second son, Johann
Heinrich (born 25 April 1736), of whom he was "dotingly fond",[31] and years had
been lost from the musical education of his other sons Jacob (born 20 November
1734) and William (15 November 1738). Isaac resolved to quit the army and find
musical employment as a civilian. 'Dismission' from military service was sometimes
hard to obtain, sometimes easy; on this occasion it was easy.

But what was he now to do? Outside of the band, musicians in Hanover were
either church organists — and Isaac could not play the organ — or members of
the Court orchestra. Each of the two dozen or so places[32] in the orchestra offered
security, congenial working conditions, and entry into the higher echelons of
society — a tempting prospect for an army bandsman sick of the uncertainties
and discomforts of military life. From time to time a particularly able bandsman

succeeded in progressing from the band to the orchestra, and Jacob was to be one of these. But Isaac was largely self-taught and his abilities did not extend this far; in later life he would make the attempt, but without success.[33]

However, he had no roots in Hanover and no compulsion to remain there — Anna, accustomed to settled farming communities,[34] said "she might have imagined her husband had dropt from the clouds"[35] — and perhaps there would be better prospects in a city the size of Hamburg. Isaac accordingly took the bold step of uprooting his family and moving to Altona, on the Elbe immediately west of the port.

It was a false move: Hamburg proved to be a city without a soul. But Isaac chanced to meet up there with a former pupil, General Wangenheim, who persuaded him that there were good prospects of peace and that he could therefore resume his place in the army band in Hanover in the confident expectation of being able to live at home with his family. Jacob, who was just fourteen, was also to have a place in the band, and William would be auditioned for a similar position when he had left school.[36]

It was an offer Isaac felt he could not refuse, and he accordingly moved his family back to Hanover and rejoined the Guards. For a time all was well. The Herschel boys, when they got home of an evening, would find their remarkable father eager to debate philosophy with them,[37] or to teach them the violin or oboe.[38] Jacob, the most talented musician of a talented family, found fulfilment in his music, but William's interests ranged more widely. He attended the private classes in Latin and arithmetic given by the teacher at the Garrison school, and after the brothers had joined the band some of their wages went towards lessons in French.[39]

The girls, on the other hand, found themselves at the mercy of an illiterate Hausfrau. Anna permitted her eldest daughter, Sophia (born 12 April 1733), to have an education of sorts and — more importantly — instruction "in all branches of needlework so as to enable her to get a livelyhood".[40] But her mother did not look back on this experiment as a success, and determined not to make the same mistake again. Sophia's skills in needlework should have made her financially independent. Instead, even after she was married, she continued to look to her parents for help in times of difficulty, when she faced the expense of moving home, or when her children needed clothing.[41]

Sophia was nearly 17 years old when Caroline Lucretia was born on 16 March 1750, and by the time Caroline was of an age to be conscious of her surroundings Sophia was living with a family in Braunschweig, presumably as a lady's maid.[42] To the best of her knowledge, Caroline first saw Sophia when her sister returned to prepare for her wedding,[43] to a fellow-bandsman of her father's, Johann Heinrich Griesbach.[44] Sophia was much loved in the family, and her mother worked all hours to spin the linen she would need in her new home. Isaac and Anna supplied her with "a gentil and comfortable outfit",[45] while her two oldest brothers took two months' pay in advance to provide the wedding entertainment, "without which in those days it would have been scandalous to get married".[46]

Her father however had misgivings about Griesbach, whom he knew to be a second-rate musician and feared would become a third-rate son-in-law. Events were to prove him right. The courtship had been conducted by correspondence, and it later emerged that Griesbach was not the author of the letters he had sent to Sophia; he had paid a friend to compose these charming *billets doux* on his behalf.[47] Griesbach's reputation in the family was to go from bad to worse. He would be remembered as "selfish, cruel and extravagent",[48] "of a selfish & quarrelsome disposition",[49] and his untimely death, when he was hardly past forty, was later ascribed to his having "shortened his days by excess of drinking".[50]

The next two daughters had died before Caroline was born, and this left her as the sole remaining supply of cheap labour in the home. Anna determined from the start to do all in her power to block her daughter's escape routes.[51] Fortune seemed on Anna's side: Caroline was not much more than 4ft 9inches tall[52] and plain in appearance, and an attack of smallpox when she was four had left her "totally disfigured" and damaged in the left eye.[53] Even her loving father was to warn her

> against all thoughts of marrying, saying as I was neither hansom nor rich, it was not likely that anyone would make me an offer, till perhaps, when far advanced in life, some old man might take me for my good qualities.[54]

This well-intentioned but grim advice left a post as governess as one of the few careers open to her. But governesses were expected to have a knowledge of French, and so her mother simply vetoed Caroline's attendance at language classes.[55]

Certain skills she was allowed to acquire, but only because they would be useful within the family. As a small girl, after attending Garrison school until three in the afternoon, Caroline had lessons in knitting.[56] Later she was taught to make household linen,[57] and "to learn grafting and washing silk stockings by way of enabling me to do this tedious kind of work which otherwise must have been put out".[58] But for lessons in fancy-work she was to depend on a kindly neighbour who during the summer of 1766 met her in secret at daybreak, for Caroline had to be home by seven ready to begin her household chores.[59] As a young adult, at a time when there were fewer members of the family living at home than usual,

Caroline would be grudgingly allowed a few months of further lessons in milli-nery and dressmaking, but on the strict condition that this was to enable her to make her own clothes and nothing more.[60] She proved popular with her teacher, who encouraged her to return for further lessons. But she could not take up the offer; her mother required her for housework, and was "very averse to my being anywhere but home".[61]

Her father wanted to teach her the violin, but this was possible only when his wife was out of the way or in a particularly indulgent mood.[62] Her four surviving brothers all became musicians of distinction, versatile performers who seemed able to play to professional standard any instrument they laid hands on. Her mother, "wishing me not to know more than was necessary for being useful in the family",[63] ensured that as a violinist Caroline would remain an amateur of modest ability.[64] Her daughter would be a household drudge forever if Anna had her way.

After Caroline, Isaac and Anna had two more children, both boys. Frantz (born 5 February 1752) died two years later in the smallpox outbreak that left Caroline disfigured,[65] and — like the siblings who had predeceased him — he is all but expunged from the family record; the Herschels were adept at responding to a setback by picking themselves up and getting on with their lives. Dietrich (born 13 September 1755) survived, a baby brother who would have a special place in Caroline's affections.

The family joy at the birth of Dietrich was overshadowed by the warclouds that were gathering once more. And this time the impact on the family would be profound, for with Sophia newly married to Griesbach, all four men of the family — Caroline's father, her two oldest brothers, and her brother-in-law — were bandsmen. Hanover and Great Britain were under a single crown; the threat from the French seemed greatest against England, and the Hanoverian Guards were ordered there as reinforcements. Sophia and her husband had made their home in an apartment in the same house as her parents' family,[66] but she now returned to her mother, bringing her furniture with her.[67]

The day for the Guards' departure came not many weeks after the birth of Diet-rich.[68] Caroline was only a child, but the event lived in her memory.

> One morning the whole Town was in motion with drums beating to march. In our room all was mute, but in hurried action; my dear Father was thin and pale and my brother Wilhelm [William] almost equally so, for he was of a deli-cate constitution and just then growing very fast. Of my brother Jacob I only remember his stating difficulties at everything which was done for him, as my Father was busy to see that they were equipt with the necessaries for a march.

The troops halooed and roared in the streets, the drums beat louder, Griesbach came to join my Father and brothers, and in a moment they were all gone.

My Sister fled to her own room. Alexander was gone with many others to follow their relatives for some miles to take a last look. I found myself now with my Mother alone in a room all in confusion, in one corner of which my little brother Dietrich lay in his cradle; my tears flowed like my Mother's but neither of us could speak. I snatch't a large handkerchief of my Father's from a chair and took a stool to place it at my Mother's feet, on which I sat down, and put into her hands one corner of the handkerchief, reserving the opposite one for myself; this little action actually drew a momentary smile into her face. I could go on describing what passed every succeeding hour throughout that day, but one word will serve for a thousand, which is, we were completely wretched.[69]

Isaac had arranged for Anna to have half his pay during his absence, and half that of the boys,[70] but Sophia's husband simply assumed her parents would be willing to support her and kept his pay for himself.[71] Caroline found herself living in an overcrowded and underfunded home, and she soon discovered that Sophia had no patience with younger children.[72] Caroline would be sent out to play with Alexander, and he would simply ignore her. In a characteristically bleak recollection of her life in the absence of her father and her favourite brother William, she was to write: "In short, there was no one who cared much about me", on reflection amending "much" to the still bleaker "anything".[73] Her childhood was not a happy one.

Caroline's mistaken recollection was that the Guards took ship for England immediately after Christmas (in the reformed Gregorian calendar) and so enjoyed a second Christmas on their arrival in a country that still followed the Julian calendar.[74] In fact, it was not until 23 March 1756 that the British Parliament debated a message from the King informing them that he had summoned a contingent of Hessian troops.[75] A motion was passed praying the King to bring over the Hanoverians, and William records that he and his comrades embarked for England at the end of the month.[76] While abroad, William learned English, and he scraped enough pennies together to buy a copy of Locke's *On human understanding*, an unusual purchase for a seventeen-year-old.[77] But Isaac's was an unusual family: for years William would write to Jacob, his soul-mate, letters of enormous length on all manner of religious and philosophical topics, and Jacob would faithfully preserve them.[78]

Even before the move to England had opened Jacob's eyes to the realities of army life, he had been hoping to obtain a position in the Court orchestra at Hanover. For this he needed his discharge from the army band, and he had been awaiting the necessary document right up to the moment that the Guards left the town.[79]

Indeed, his failure to get it had cast a shadow over the celebrations at the birth of Dietrich.[80] Worse was to come. While the Guards were in England, Jacob learned that someone else had secured the appointment he coveted, and no sooner had the news reached him than his request for a discharge was granted.[81] In the autumn of 1756 he returned home in some style as a civilian, Isaac and William following on foot with the regiment at the end of the year.[82]

At the hour when the Guards were due back in Hanover, Caroline's mother was hard at work with the cooking, and so she sent Caroline out to look out for her father and brother. Failing to find them, Caroline returned disconsolate to the apartment, only to discover that they had already arrived and a party was in progress. In a gesture she always remembered with gratitude, William it was who left the celebrations to welcome and comfort his little sister.[83] Perhaps she even forgave him for the time he had been put in charge of her, and had buried her upright in sand as the only way to keep her quiet.[84]

Her father and favourite brother were now home, and Sophia had returned to her own apartment;[85] but this happy state of affairs did not last, for on 1 May 1757 the Guards marched out of Hanover once more, and this time it was to war. Isaac, William and Griesbach were in the band as before, but the fortunate Jacob was no longer a bandsman and so he remained at home.[86] William writes:

> my Father and I went with the regiment into a campaign which proved very harassing, by many forced marches and bad accommodations. We were many times obliged, after a fatiguing day, to erect our tents in a ploughed field, the furrows of which were full of water.[87]

Worse, much worse was to follow. On 26 July, the Hanoverians and their allies fought the French at the Battle of Hastenbeck. Fortunately for Isaac and his son, they were non-combatants. During the engagement, William found he possessed an enviable ability to close his mind to the carnage going on around him, if we are to believe what he was to tell his son many years afterwards: "with balls flying over his head he walked behind a hedge spouting speeches, rhetoric then being his favourite study."[88] In the aftermath of the battle his father "advised me to look to my own safety",[89] so William set out for home, only to find on arrival in Hanover that he risked "being pressed for a soldier" in the makeshift corps being desperately assembled for the defence of the town. He and his mother decided it would be safer for him to retrace his steps and rejoin the army, where he would be a non-combatant.[90] Before long he met up with his regiment, where he not surprisingly "found that no body had time to look after the musicians; they did not seem to be wanted".[91]

The outcome of the battle had been even more confused than is usually the case: both commanders thought they had been defeated, but unfortunately for the Hanoverians it was the French commander who first realized his mistake.[92] There then followed weeks of uncertainty and danger. William tells us that "The weather was uncommonly hot and the continual marches were very harassing".[93] Isaac was beside himself with worry for William; and not only for William but for Jacob, who was at risk from being pressed into the defence of Hanover. Even after the French entered the town on 10 August and any attempt by the inhabitants to oppose them ceased,[94] Jacob was in the uncomfortable position of an able-bodied man in territory occupied by the enemy. Eventually Isaac

> resolved to let them go to England, where he doubted not they might through the recommendation of Mr Barband (and some other acquaintances they had had an opportunity of making when in England before) find some engagement as Teachers or Performers.[95]

Jacob of course was under no obligation to stay. Nor, it seemed, was William: as his father saw it, Isaac himself was a grown man under oath who therefore had no choice but to remain with the regiment, but William had not yet made this commitment. Isaac was confident he could obtain his son's formal discharge when normality returned, as indeed he did.[96]

The next skirmish added urgency to Isaac's advice,

> for when hearing the balls whistle over their heads through the branches of the tree under which they had taken shelter, my Father cried out Go! Go! waving his hand towards Hamburg Go! You have no business here![97]

We may suppose that Isaac's actual words were a good deal less formal than those reported by Caroline: perhaps the German equivalent of "Why don't you get the hell out of here?". William, finding that "no body seemed to mind whether the Musicians were present or absent", set off for Hamburg, as a staging post for England.[98] If Caroline is to be believed, Isaac was put briefly under arrest for conniving at William's departure.[99]

The allied commander was the Duke of Cumberland, one of the sons of George II, Elector (in effect, King) of Hanover and King of Great Britain. In the aftermath of Hastenbeck, Cumberland had withdrawn his troops northwards to Stade, on the left bank of the Elbe, below Hamburg, thereby leaving almost the whole of the Hanoverian region in the hands of the French. A truce was reached on 8 September at the Convention of Klosterzeven, better described by its alternative title: the

Capitulation of Klosterzeven. Cumberland agreed to terms whereby the allies of the Hanoverians would return home, but the Hanoverian troops — most of whose homeland was now occupied by the enemy — were to remain near Stade. They would thus be forced to stand by helpless while the town of Hanover and the surrounding region were at the mercy of French occupying forces. Their families went hungry while the countryside was plundered. Enemy officers, together with some sixteen soldiers, would be quartered in the very house where Anna and her beleaguered family had their apartment.[100]

Jacob was hiding in the apartment, as he had been since Hastenbeck. The Herschels were conscientious to a fault when it came to working hard, but their civic instincts were less developed, and Jacob had no intention of becoming a hero. To dodge first the "hot press for soldiers" by the Hanoverian authorities trying to defend the town,[101] and later awkward questions from the the occupying French troops,[102] Jacob "was obliged to live in concielment in our discomfortable family".[103] He emerged only briefly, at the Garrison Chapel on 13 October, to stand as godfather to Sophia's firstborn.[104] Soon he too slipped out of town, and made his way to Hamburg where William was waiting. Together they took ship for England, fortunate that a talented musician can earn a living in any land.

William had asked their mother to forward a box of his things to Hamburg. He particularly wanted his books, and some celestial globes he had made himself. But the illiterate woman had little patience with such trinkets. "She was several days and throughout the nights busied with ironing and drying linnen by heated stoves",[105] but the books and globes she gave to Caroline and her baby brother Dietrich as playthings, and before long they were in pieces.[106]

Meanwhile, both sides in the conflict looked for reasons to repudiate the Convention,[107] and hostilities resumed. Cumberland had been replaced by the able Prince Ferdinand of Brunswick, and the Hanoverians and their allies rallied under their new general. In mid-February 1758 anxiety in the mind of the French commander gave way to alarm when rumour reached him of the impending arrival of a British expeditionary force. By the 24th he had hastily gathered his scattered forces near Hanover and begun the march south.[108] It seems that when the French army abandoned the town, they threw the food that they could not carry away into the river,[109] leaving the inhabitants starving, shivering for want of wood from the forests that the French had cut down,[110] and with typhus raging. Later, in 1761, Caroline herself was to become became a victim of the disease;[111] at one time her life was despaired of, and the typhus left her so weakened that "for several months after I was obliged to mount the stairs on my hands & feet like an infant".[112]

Ferdinand's forces, of which Isaac was the humblest of members, prospered, and by the end of the year the French had been temporarily expelled from the Hanoverian region. But there was no let-up in the campaigning, and as yet no hope of a return home for Isaac and his comrades. Meanwhile Anna somehow managed to keep three children fed, clothed and warm on the money her husband was able to send her,[113] supplemented by needlework put her way by a friend who supplied the army with tents and linen.[114] Little Caroline helped as best she could, but she found it painful work as she had not yet learned to use a thimble.[115] Anna was a resourceful woman, although there was a price to be paid for this. Even in normal times, "my Mother's way of treating children was rather severe",[116] and these times were far from normal.

In some ways Caroline was affected more than anyone by what had happened, because the two members of the family who showed her affection were gone: her father caught up in a war that seemed to have no end, William abroad and likely to stay there forever. But child though she was, she had a crucial role to play as scribe to the neighbourhood, for her mother was not the only wife incapable of writing to her absent husband.[117]

On 7 January 1758 Sophia's husband had returned to Hanover, having been appointed Town Musician at nearby Coppenbrügge.[118] Why he was free to come and go in this way is a mystery, but so it was. He and Sophia accordingly vacated their apartment and moved to Coppenbrügge.

They took Alexander with them, as he was to serve an apprenticeship there in a Guild of Music.[119] But Griesbach proved to be a "brute"[120] who saw the boy as a source of cheap labour. Alexander had to chop the firewood, dig the garden, and mind the baby.[121] Not only that, but when Anna went to Coppenbrügge to assist Sophia in her second confinement, she witnessed Griesbach give Alexander a box on the ear, without cause and despite the boy's suffering from mumps at the time. Only after Griesbach promised to reform would Anna leave for home without taking her son with her.[122] Alexander remained in Coppenbrügge to complete his lengthy apprenticeship, and to enhance further his popularity with the local girls and the landlord of the tavern,[123] after which he would return to Hanover with appointment to the sinecure of Town Musician.[124]

We are poorly informed on life in the Herschel home in the months following the French evacuation of Hanover, because Caroline characteristically destroyed the account that later she wrote of this difficult time. Intending her narrative for the eyes of Dietrich, she says she will limit herself to remarks that will allow him to see how vainly his poor Sister has been struggling throughout her whole life, for acquiring a little knowledge and a few accomplishments; as might have saved her from wasting her time in the performance of such drudgeries and laborious works as her good Father never intended to see her grow up for.[125]

Her gloomy remarks, we note, are not limited to her childhood but embrace "her whole life", including therefore the half-century she was to spend in England with her brother William.

At length, in July 1759, Ferdinand established his forces at Hille, some forty miles west of Hanover, and despite the distance, and the presence of French forces to the south, Anna and other plucky wives set out to visit the husbands they had not seen for nearly two years.[126] Anna returned home from her four-day stay appalled at the conditions in which Isaac was living and the harm this was doing to the asthmatic condition from which he had suffered ever since the Battle of Dettingen.[127]

On 1 August 1759, at the Battle of Minden, not far from Hille, Ferdinand won a victory so complete that Hanover was never threatened again. Isaac and Anna were agreed that Isaac should urgently seek a discharge before another campaign could begin, but the formalities took time, and it was not until 20 May of the following year that Isaac at last arrived home to be greeted by his "helpless and distracted family".[128]

William for his part preferred to stay in England. After a difficult period during which he had come close to abandoning music,[129] his career was at last flourishing; certainly, after his experiences of army life, he had no wish to resume his place in the band of the Guards. Jacob was less content with life abroad. He had a justifiably high opinion of his abilities as a musician, and aspired to the status he thought appropriate. Accordingly, he considered it a "degradation" to perform unless he was himself first violin, and simply refused to do so, even if this meant going hungry. The long-suffering William — who was prepared to take a similar stand on occasion himself[130] — found himself helping to support his older brother.[131] Jacob's strategy evidently paid off, for after the final liberation of the region around Hanover he was provisionally offered the post of first violin in the Court orchestra there.[132] But in order to return home for the necessary audition[133] he had once more to ask William for help, his younger brother "parting with his last farthing for travelling expenses".[134]

The audition successfully negotiated, in January 1760 Jacob found himself employed in an official residence of the reigning sovereign, and entitled to entry into all strata of society — with the exception of the nobility, who formed a class apart. As a 'titled person', he "held his head very high, and expected his mother and sister to feel honoured by waiting on him".[135] He was also the principal wage-earner in the family, and so could dictate where the family lived. The Herschels resided in a succession of the flats into which the tall old houses of Hanover were divided,[136] and they seem to have been forever on the move. Once at least the move was fully justified — a forge was newly installed underneath them, making the home no place for musicians needing to practice[137] — but the move they made on Jacob's return from England was solely because he thought his old room no longer grand enough. He took another apartment some distance away, arranging for his mother and siblings to move into the same house as soon as there was a vacancy.[138] Meanwhile he would regularly dine at his mother's, and

> ... it generally happened that before he departed his Mother was as much out of humour with him as he was at the beefsteaks being hard, and that I did not know how to clean knives and forks with brickdust. This extreme hurry of introducing English cookery &c. into our frugal household all at once, had a great effect on my mother's temper. She became exceedingly passionate, and generally I had the ill luck of falling under her displeasure through my awkwardness in assisting her.[139]

Jacob too vented his displeasure on poor Caroline, who "got many a wipping for being too awkward at supplying the place of a footman or waiter".[140]

Caroline's account of the events that followed their father's return the following May, written in 1821 as William's life was approaching its close, is condensed and characteristically gloomy,

> for I fear — in my perfectly isolated situation — I shall want spirits, as well as time, for living over again a laborious life of a succession of disappointments.[141]

At first Isaac's return brought about a long-overdue improvement in the family fortunes, although the parents soon became increasingly worried about the behaviour of their highly talented eldest son. If Caroline is to be believed, Jacob was proud, vain and a spendthrift, "whose whole attention was engrossed by two lady scholars for upwards of 4 years".[142] But the Herschel talent for music was already manifesting itself in little Dietrich. Isaac had not been home more than a couple of hours before he had restrung a small violin and given the four-year-old what was to be the first of his daily lessons.[143] On such occasions Caroline would sit in the corner knitting "and listening all the while", and sometimes Isaac would be able to give her too a lesson, after he had finished with Dietrich. Eventually she progressed enough to play the least challenging of the violin parts in gatherings of pupils and friends.[144]

But her mother's demands for help in the kitchen were unrelenting:

> ... sometimes I found it rather too much for getting through the work required, and felt very unhappy that no time at all was left for improving myself in music or fancy-works [in which she had been having secret lessons].... Though I had neither time nor means for producing anything [in fancy-work] immediately either for shew or use, I was content with keeping samples of all possible patterns in needlework, beads, bugles, horsehair, &c. For I could not help feeling troubled sometimes about my future destiny in case I should loose my dear Father, and my Brother's getting married; for I could not bear the idea of being turned into an Abigail or housemaid, and thought that with the above and such like acquirements with a little notion of Music, I might obtain a place as governess in some family where the want of a knowledge of French would be no objection.[145]

Meanwhile life in the Herschel household was fraught. The war rumbled distantly on, and Sophia's husband thought his family vulnerable in a village that lacked any means of defence. As a result, from time to time, the Herschel home would be invaded by Sophia complete with children, their furniture and even that of their neighbours.

> And I think to such like harassing disturbances from which my Father never found respite but in his grave, I must ascribe not only the disagreeable feeling of dependance to which his younger children for many years were doomed; but the mortifications and disappointments which have attended me throughout a long life.[146]

The foundations of Caroline's gloomy and often resentful outlook were being laid, and who shall blame her?

The final parting from her beloved father, the only person in her life who never ever failed her, was soon to happen. The hardships of army service had taken a severe toll of Isaac's health, and he found it an ever-increasing struggle to give his music lessons and to do what little else he could to provide for his family. And meanwhile his attempts to give Caroline "something like a polished education" were being thwarted by Anna, who "was particularly bent upon it, it should be rough, but at the same time a useful one".[147] In August 1764 Isaac suffered a paralytic seizure that left him temporarily speechless and deaf.[148] He could no longer perform on the violin, but his loyal scholars stood by him. He struggled on, copying music while Caroline read to him, and giving lessons as best he could.

His final months did however bring one consolation. Jacob's current lady-love

Miniature (in a private collection) of William as a young man. The artist and date are unknown, but it seems likely that the miniature (height 47mm) was given to Caroline by William when he visited Hanover in 1664. In the lid a note by one of William's descendents states that it was given by Caroline to William's son John, presumably on one of his visits to his aged aunt after she had retired to live once more in Hanover.

for some reason demanded that he leave Hanover with her; this would have meant giving up his exalted place in the Court orchestra, a sacrifice that Jacob refused to make, and to the relief of his parents their engagement was terminated.[149]

At the end of February 1767, Isaac was confined to bed. Honourable to the last, and despite the poverty that faced his wife and younger children, he would not agree to a fraud involving a form of life insurance, for which his physician offered to supply him with a spurious certificate of health.[150] Three weeks later, on 22 March, he died.[151]

Caroline was devasted, first by her bereavement, and then by the realization of her predicament. Now aged seventeen, she was at the mercy of her mother and her eldest brother Jacob, who was the new head of the family. Even before their father's death Jacob had been "incorrigible where Luxury, ease and Ostentation were in the case",[152] running up bills for clothes and entertainment that exceeded the wages he brought home. Now he was free to indulge himself with even less restraint. He and their mother "seldom were satisfied with each other";[153] Jacob, a talented musician at ease in the presence of royalty, was probably irritated and embarrassed by the illiterate Anna.

For Caroline the outlook was bleak indeed. Thus far she had been a teenage daughter living with her parents and earning her board and lodging by the help she gave her mother. Now, with her father dead, she was dependent upon Jacob; and it was far from clear what would become of her when he got married.

William had paid a visit to Hanover in April 1764, when he saw his father for the last time and could allay his anxieties about his son's prospects in England.[154] The visit lasted less than two weeks,[155] and by ill luck it coincided with Caroline's confirmation. Their mother set great store by confirmation, even agreeing — though grudgingly — to give Caroline time off work to attend the necessary classes. Her regular schooling, these classes, and "doing the drudgery of the scullery"[156] meant that for much of the visit of her favourite brother she was barred from spending time with him. The final straw came on the following Sunday when she was in church for her first communion: the coach carrying William away passed the open door, and the postillion chose that moment to give "a smettering blast" on his horn.[157]

Late in 1766 William took up the post of organist to the Octagon Chapel in the fashionable spa town of Bath, where there were many opportunities for enterprising musicians during the winter season; and in the years that lay ahead there was much fraternal coming and going across the English Channel. Jacob was the first to arrive in England, in June 1767, and would stay for two years. The focus of Court life was in London and Windsor rather than Hanover, and besides, in Bath William could secure him profitable engagements. In the classical pattern of patronage, Jacob dedicated a set of sonatas to the Queen, and was duly summoned to Court. He performed so well there that his annual salary was increased by 100 thalers, with expectation of still better things to come.[158]

Their father on his deathbed had committed to Jacob the musical education of Dietrich,[159] who was eleven years old at the time; but as a result of Jacob's absence abroad this had passed into the questionable care of Alexander, who with his apprenticeship completed was back home once more. However, after some months in England Jacob's conscience pricked him, and he summoned Dietrich to Bath.[160] But their mother had grave misgivings at permitting this while the lad was still too young to be confirmed, misgivings that were only partially allayed by assurances that he would be confirmed in England; and when she discovered after some months that this had not been done she demanded that Jacob bring him back to Hanover, which he did.[161]

Jacob visited England briefly a second time in 1770, on this occasion taking with him Alexander, who was to settle in Bath, remaining there until 1816 when he retired to Hanover. The international character of music, and the remarkable musical talent and versatility of every one of the brothers,[162] ensured that they could support themselves equally well in either country.

One result of this traffic was that William was kept informed of developments in Hanover, and of the plight of his little sister. For Caroline, yet more vexation had followed Jacob's return to Hanover in the summer of 1769. "So many luxurious fashions all at once were introduced"[163] that their mother was forced to hire a servant girl to help Caroline: good news, except that the two girls had to share the same bed, which Caroline resented.

Meanwhile their mother and Jacob were forever quarrelling, and Caroline began to worry more and more about what was to become of her:

> I began to feel great anxiety about my future destination, for I saw that all my exertions would not save me from becoming a burden to my brothers, and I had by this time imbibed too much pride for submitting to take a place as Ladiesmaid, and for a Governess I was not qualified for want of knowledge in languages.[164]

Looking back in old age, she spoke of Hanover as "where the first twenty-two years of my life (from my eighth year on) had been sacrificed to the service of my family under the utmost self-privation without the least prospect or hope of future reward".[165]

Finally, early in the winter of 1771/72, and no doubt with the full approval of Alexander, William decided to act.[166] He wrote to his mother and Jacob, and

> proposed my coming to Bath, where by his instruction I might soon become a usefull member to the musical profession as a singer or teacher, having heard from Alexander that I had some notion of music and a good voice.[167]

Caroline, we might think, could hardly teach what she did not herself know; and the odds against this Hanoverian household drudge who had never left home having the ability (and confidence) to stand on a platform before an audience of English high society and deliver the soprano solos of Handel's *Messiah* were truly daunting. The possible career in music must have been an excuse rather than a serious prospect. As William was unmarried she would have more than earned her keep as his trusted and hardworking housekeeper, the female head of his household.

Jacob, who "of course never heard the sound of my voice but in speaking",[168] was at first taken by the idea, and seemed willing to give his consent. But then the absurdity of the proposal struck him; he turned to ridicule,[169] "and made himself very merry".[170] For her part, Caroline hardly dared hope against hope, but she would do anything within her power to advance William's plans. She knew that novice singers were advised to practise "by singing with a gag between their front teeth", and so whenever the apartment was empty she would retire to a back room, close the windows, get some needlework, put a gag between her teeth, and begin to imitate the solo passages of concertos, "shake and all".[171] But she felt guilty about the prospect of leaving her family, and so set about making them a stockpile of clothes. "For my Mother and brother Dietrich I knitted as many cotton stockings as were to last two years at least."[172] She made net ruffles; if she stayed, these would be for William as a token of thanks for his efforts on her behalf, while if she left, Jacob could have them as a parting gift.[173]

William arrived on 2 August 1772. As luck would have it, the Court orchestra was then away from Hanover, attending on the Queen of Denmark at the royal hunting lodge at Göhrde, near the Elbe some eighty miles northeast of Hanover.[174] Dietrich, who was deputizing for the absent Alexander, had just returned from this engagement,[175] but Jacob was still away. As head of the family he was anxious to be involved in the decision over Caroline's future, but he was the first violin of the orchestra and unable to obtain leave. He could do no more than vent his frustration in letters that "expressed nothing but regret and impatience at being thus disappointed".[176]

The time for decision had come, for William's commitments in Bath made it imperative that his stay in Hanover last no more than a fortnight.[177] Their mother's objections had been easily overcome, for William knew exactly where her priorities lay:

> ... my Mother having consented to my going with him, and the anguish at my leaving her being somewhat alleviated by my brother's settling a small annuity on her, by which she would be enabled to keep a person for supplying my place.[178]

Would the history of astronomy have been different if Jacob had been at home and party to the discussions? It is hard to say. He and William had always been on excellent terms, and one might suppose that Jacob could hardly hold Caroline a prisoner against her will. She was after all a grown woman of twenty-two. Yet she saw herself as requiring his permission, and was "uncomfortable" at going "without taking leave or obtaining his consent for doing so".[179]

Very soon William would have to depart, with or without his sister. Days passed, and still Jacob did not appear. Finally, Caroline made the fateful decision: "without being able to effect a meeting, I was at last obliged to [go] without taking the consent to my going from my eldest brother along with me."[180]

Life for Caroline the household drudge was about to be transformed. "It was my lot", she later wrote, "to be the Cinderella of the family".[181] Cinderella was now going to the ball.

2

Musical Interlude

On 16 August 1772 Dietrich accompanied their mother Anna to the Posthouse in Hanover, to wave goodbye to William and Caroline as they set off in an open coach.[1] Caroline would never see Anna or her sister Sophia again; and only when half a century had passed would she return to Hanover, following William's death, to share a home with Dietrich once more.

After six days and nights of travel in high winds, and with Caroline's hat lost in a canal,[2] they reached the little port of Helvet. From there they were ferried out in stormy seas to where the packet boat lay at anchor. The crossing would live in Caroline's memory, and with good reason: when the battered ship finally limped to an anchorage off the coast of East Anglia, it had lost its mainmast and a second mast as well. They covered the last perilous yards of their journey to England in a small open boat, "to be set, or rather thrown like balls" on the shore.[3]

They made their way to nearby houses, where they found some of their ravenous fellow passengers already breakfasting from fine wheaten loaves which serving women were cutting as fast as they could. Revived with this and cups of tea, and after Caroline had changed her clothes, she and William were taken by cart to the road where stage-coaches or 'diligences' passed on their way to London. Unfortunately, the coach that stopped to pick them up was being pulled by a horse that was unused to working between shafts, and it promptly bolted. Brother and sister, along with a third passenger, threw themselves out, Caroline ending up in a ditch which fortunately proved to be dry. A gentleman and his servant, who by lucky chance had chosen to accompany the coach on horseback, came to their rescue; and to prevent any repetition of the incident, their two benefactors escorted the coach all the rest of the way to London, where they arrived at midday.[4]

There they were to stay the night in a city inn. William had business to attend to, but in the evening he took Caroline to see the sights: St Paul's Cathedral, the Bank, and — an omen of what lay ahead for her — opticians' shops. The following evening

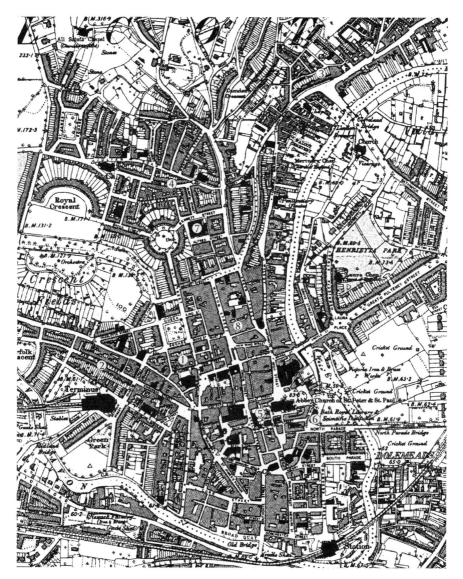

A nineteenth-century map of Bath, showing places associated with William and Caroline. Based on information supplied by Richard Phillips. (1) Beauford Square, where William occupied two different houses between March 1767 and June 1769; (2) New King Street, where William occupied no. 7 from June 1769 to midsummer 1774, and no. 19 from September 1777 to December 1779 and from March 1781 until July 1782; (3) Walton Parade, William's home from midsummer 1774 to September 1777; (4) Rivers Street, where William occupied no. 27 from December 1779 to March 1781, and where the Bath Philosophical Society was established; (5) the Pump Room; (6) the (Old) Assembly Rooms; (7) the New Assembly Rooms; (8) the Octagon Chapel. The Turnpike, said by Caroline to have been near their home at Walton, is half a mile northeast of Walton Parade.

they made their way to another inn, to catch the ten o'clock night coach to Bath. There they arrived at four the next afternoon, 27 August, "almost annihilated".[5]

At William's home, 7 New King Street, they were greeted by an elderly lady, Mrs Bulman, and her daughter. Prior to his move to Bath in December 1766 to take up the post of organist at the Octagon Chapel, William had lived in Leeds with the Bulmans,[6] but when he parted from them the Bulmans were in dire straights because Mr Bulman's business had failed. On first arrival in Bath William had lodged with the Harper family in Bell Lane. Soon he found himself enchanted by their daughter Elizabeth; Elizabeth sang as she sewed, and in no time the resourceful William was giving her lessons. However, "on disclosing his passion he received no encouragement",[7] and so he began to look around for a home of his own. The arrangements with the Bulmans had suited him well, and the post of Clerk to the Octagon Chapel was not yet filled. By March William was able to write to Mr Bulman with the offer of the post and the invitation to share accommodation once more. On 25 March, Lady Day, he took a lease on a house in Beauford Square; he moved in on 6 April, and on the 16th the Bulmans arrived to join him. Since then William and the Bulmans had moved house more than once, but always together, and now they were installed in New King Street. The rent was 30 guineas a year,[8] of which the Bulmans were paying one-third, and Mrs Bulman managed the establishment.

When Caroline joined him, William and the Bulmans had happily coexisted in Bath for over five years, and William was understandably anxious that the arrival of his sister should not disturb these long-standing arrangements. During the journey, therefore, he had "taken every opportunity of making me hope to find in Mrs. Bulman a well-informed and well-meaning Friend, and in her Daughter (a few years younger than myself) an agreeable companion".[9] William's sister, as he well knew, was long on honesty and short on tact. Caroline was to tolerate Mrs Bulman's kind attempts to enlarge her culinary skills, but the daughter she thought "little better than an idiot".[10]

The Bulmans occupied the ground floor of the house, and William the first floor, the front room of which was "furnished in the newest and most handsomest stile"[11] and spacious enough for rehearsals and chamber music. The attic was to be shared between Caroline and their brother Alexander, who had been living with William for the past two years. On first arrival Caroline had hoped that Alexander would be present to welcome her, a familiar face speaking a familiar language, but he was away, and would not return before the beginning of the season. Caroline accordingly "found myself all at once in a strange Country and among straingers".[12]

Having only twice had a bed to sleep in since leaving Hanover, Caroline did not wake until the following afternoon. The time allowed her for relaxation was brief. Next morning at their seven o'clock breakfast William

> began immediately giving me a lesson in English and arithmetic, and showing me the way of Booking and keeping account of cash received and laid out. The remainder of the forenoon was chiefly spent at the harpsichord, shewing me the way how to practice singing with a gag in my mouth.[13]

The season at Bath would not begin until October, and so "my Brother had leisure to try my abilities of making a useful singer for his concerts & oratorios of me, and being very well satisfyed with my voice I had 2 or 3 lessons every day".[14] But Caroline was discovering that William's commitment to music was now less than total. "And by way of relaxation we talked of Astronomy and the fine constellations with whom I had made acquaintance ... travelling through Holland."[15]

Before long Caroline had enough English to speak with Mrs Bulman and "the hotheaded old Welsh woman" that William kept as a servant. Alexander was now back, although he and William were preoccupied with the start of the season. Each Sunday Caroline would be issued with her weekly allowance of cash for the household expenses, and at the end of the week this would have to equal the outgoings in her housekeeping book plus the sum remaining in her purse. In time she plucked up courage to venture forth to the shops and market, "and brought home whatever in my fright I could pick up", unaware that Alexander was shadowing her, ready to come to the rescue if she got into difficulties.[16] Family legend was to insist that on one occasion by mistake she brought home a live sucking pig.[17] This story may be apocryphal; but she herself admits that later, in London, she selected two horses for purchase, both of which later turned out to be blind.[18]

Caroline was responsible for the servants, whom she no doubt expected to work as hard as she herself had had to do back home in Hanover, and she invariably brought out the worst in them. On arrival in Bath she had been appalled to find the ivory handles of the knives and forks were damaged and the blades rusty, but William was otherwise preoccupied and she had to deal with the situation herself as best she could. Drawing on all her limited reserves of tact, she represented the instructions she gave the maid as originating with William, but "they were received with so much ill will and in short she gave warning and left us at Christmas".[19]

Mrs Bulman neglected to warn her to ask for a character reference before hiring a replacement; the Herschels therefore suffered from a succession of dishonest housemaids,[20] "pickpockets & streetwalkers" among them.[21] At times they were without servants altogether, "and I was vexed at being thus so pittifully interrupted in my practice".[22] Eventually a member of the chapel choir advised Caroline to speak with a maid's previous employer before hiring her.[23] Caroline's relations with the servants were never less than fraught, and this she ascribed to xenophobia on their part.[24]

Understandably, Caroline's first winter in England was not a happy one, and she struggled against "Heimwehe [homesickness] and Lowspirits".[25] She was eager to do everything in her power to foster her career as a singer, but already there were warning signs that this career might be blighted by the endless demands William was placing on her:

> I soon began to fear when I saw one care after another devolving on me that the hours required for practice would be much abridged, and that my Brother would not have much time to spare for giving me many lessons, for the town soon began to fill and except at meal times he was seldom at home.[26]

Despite her closeness to William and Alexander, her autobiography expresses what was to be a recurring regret: "I still was and remained almost throughout my long life without a friend to whom I could have turned for comfort or advice when I was surrounded by trouble and difficulties."[27] She adds, with a hint of bitterness: "This was perhaps in consequence of my very dependent situation, for I never was allowed to form any acquaintance with any other but such as were agreeable to my eldest Brother."

In January came the tragic news from Hanover that Sophia's husband had died, leaving her with six children, one an infant,[28] and William and Alexander had to pay off her debts.[29] Caroline was currently seeing little of these brothers with whom she shared her home, for in Bath the winter season was in full swing. William was occupied all hours, and Alexander had cello lessons to give as well as other musical engagements. Even when Alexander was free to spend time with his sister,

> it did me no good, for he never was of a cheerful disposition, but always looking on the dark side of everything, and I was much disheartened by his declaring it to be impossible for my Brother to teach me anything which would answer any purpose but that of making me miserable.[30]

Caroline's limited English was a major cause of her alienation from those around her. To help her make friends, William persuaded her to sing at his private parties, and to take part in the rehearsals and Sunday services of his Octagon Chapel choir.[31] The chapel was one of a number of strictly private places of worship established in Bath in the eighteenth century, where the well-to-do who came to take the waters and enjoy the social life might worship their Maker without rubbing shoulders with the lower orders. It had been paid for by a subscription organized by a banker, William Street, and the Rev. Dr John de Chair, who himself became the minister.[32] Building began in 1766, but when William had arrived at the end of the year to take up his post as organist it was not yet finished. Only in June did the installation of the organ begin,[33] and it was October before the chapel could be officially opened.[34] Later that month William directed two performances there of *Messiah*, and played one of his own concertos on the new organ between the second and third acts.[35]

William's chapel duties left him with ample time for other profitable engagements. Within days of his arrival he had promoted the first of the many 'benefit' concerts he was to give in Bath and nearby Bristol. Though thinly attended it served to announce his appearance on the scene — William displayed his versatility by performing on the violin, oboe, and harpsichord — and he was soon invited to join the salaried band that played at the Pump Room, the Assembly Rooms, and elsewhere.[36]

Benefit concerts were organized and paid for by the musician concerned, who pocketed the profits, if any; and the competition to secure the best venues, dates, and artists could be intense. On arrival William had found Thomas Linley, Sr, said by one observer to be "a very sour, ill-bred, severe, and selfish man",[37] in possession of the field. Whatever his faults as a human being, not only was Linley a fine and enterprising musician, but he had a large and talented family whose services he could call on himself — and deny to rivals. A clash with the equally enterprising William was almost inevitable, and the poor management exercised by the town's Master of Ceremonies (who administered the concerts and ensured that etiquette was observed in the dances in the Assembly Rooms[38]) did not help. Matters came to a head after the New Assembly Rooms were opened in the summer of 1771, in head-to-head competition with the Old Rooms. The Director of Music in the New Rooms was to be Linley rather than William. Not only that, but William was not among the ten musicians recruited to play at the balls on Mondays and Thursday; he was merely one of the three additional performers hired for the Wednesday concerts.[39]

By the turn of the year the normally even-tempered William had found a trivial excuse — Linley's failure to provide him with a music-stand — to storm out. Linley responded sarcastically by apologising in the local newspaper for this "grievous Wound" which "must violently agitate the tender Sensibility of his Frame". William retorted that he would make allowances for the "sensibility of Mr. Linley's *frame*", which clearly was not "tender enough to perceive the real Offence there is in leaving *any* Gentleman of the Band two *successive* Nights without a Desk". Before long Linley was writing of William's "mean and contemptible Disposition" and referred to "the Malice of a Slanderer", comments that to William revealed "that Bitterness of Temper which is the general Attendant on low Cunning and dark Envy, when they are drawn out of their lurking Place and exposed to Public View". William lost no time in organizing concerts in opposition to Linley's, and he threw in his lot with Linley's enemies who based themselves on the Old Assembly Rooms.[40]

Strife among the Bath musicians continued for years, though William was to develop more profitable ways of making money than by competition with Linley. He gradually reduced his public engagements; instead, he took great numbers of pupils, often as many as eight a day, and many of them were wealthy aristocrats. His greatest patron was the Marchioness of Lothian. In a typical season, 1775–76, this formidable lady marshalled her friends to such effect that on twenty successive Saturdays William gave private concerts in their various homes. "As the music was chiefly to consist of the singing and harpsichord playing of my scholars I engaged only a sufficient accompanyment to make up a quartetto." Minimum expenses, maximum profit.[41] Caroline would rarely be involved: she was not one of his fee-paying scholars, nor would she then have been comfortable in aristocratic society. Furthermore, William presumably rode to such engagements on horseback, and this would have been inappropriate for a lady dressed for such an occasion.

William's battles find surprisingly little echo in Caroline's recollections of her first season in Bath. She seems less concerned with the success or otherwise of William's concerts than with his absence when she wanted a lesson, as she realized to her dismay that she was far from the top in William's order of priorities. But even William had to stop for breakfast, and while he ate Caroline sang for him and he coached her.

> But by way of [my] not getting out of practice my Brother used to rehearse my lessons whilst at breakfast and preparing for going to his scholars, and at 8 o'clock I generally was left with his directions how to spend 5, 6 &c. hours at the Harpsichord.[42]

She accepted that this was the most William could do amidst the press of winter engagements, but when spring came and the season began to wind down, Caroline looked forward to "receiving a little more of my Brother's instruction ... but I was greatly disappointed".[43]

> I hoped and expected my Brother would now help me on a little more in my musical practice; and he was still pleased with my voice; but gave me no encouragement for bestowing much time in attempting to become a proficient player on the harpsichord, and that he would always acccompany my songs &c.[44]

Caroline's "expected" implies a certain entitlement. As to teaching her the harpsichord, it was uncharacteristic of William to discourage an eager pupil; but probably he realized that she had less than her share of the Herschel musical genes, and that coaching her on the keyboard would prove a waste of time. Though she acquiesced in his refusal, Caroline came to regret her inability to play an instrument so widely in demand. Another possible career had been closed to her:

> I was contented with this arrangement, but oftens repented my following his advice, for after the loss of my Capellmeister my voice became of little use to me.[45]

William had been coming home exhausted by his day's work, too weary to bother with Caroline and eager to retire early to bed "with a bason of milk, sago or a glass of water"[46] and a book on astronomy. This would lead next morning to "an astronomical Lecture of which Alexander generally was a partaker".[47] William had become intrigued with astronomy, and it was Caroline's misfortune to arrive in Bath just as his initial fascination was turning into an obsession. William would soon be leading a double life, as professional musician and amateur astronomer; and this would have its impact on those around him, Caroline in particular being "much hindered in my practice by my help being continually wanted in the execution of the various [astronomical] contrivances".[48]

Much has been made of Caroline's childhood recollections of her father's interest in astronomy: of how he had taken her out one evening to show her the constellations,[49] and of his gathering the family round a tub of water in the courtyard to observe in safety an eclipse of the Sun.[50] In retrospect these little incidents may have assumed a significance for her they did not have at the time. More of an omen for the future is her memory of William's teaching her the constellations as they journeyed across Holland in the open coach, the skies swept clear of clouds by the high winds.[51] William was taking the first step towards sharing with his sister the vocation that would eventually replace music, the quest for an understanding of the universe, in which she would one day be his collaborator.

William was currently earning nearly £400 per annum,[52] a handsome salary, and he could afford to indulge his new hobby. His diary records that in April 1773, as Caroline's first season drew to a close, he purchased a quadrant, which would enable him to measure the angular heights of heavenly bodies, and on 10 May he "Bought a book of astronomy and one of astronomical tables".[53]

The book William bought was James Ferguson's hugely successful introductory text, *Astronomy Explained Upon Sir Isaac Newton's Principles*.[54] Ferguson was the son of a Scots tenant farmer, and his schooling had lasted just three months. But he had developed into an ingenious constructor of orreries and other mechanical models of celestial movements, and his abilities had caught the attention of Professor Colin Maclaurin of Edinburgh. As a result Ferguson was now a Fellow of the Royal Society, and unofficial scientific "popularizer in residence" to the court of King George III, who had succeeded to the throne in 1760;[55] he even had a pension from the King of £50 a year.[56]

Ferguson's *Astronomy* was published in 1756, and was still appearing in new editions in the nineteenth century. It dealt not only with the solar system but with tides, the equation of time, the calculation of new and full moons and eclipses, and of course descriptions of his mechanical models. Only in the second edition did Ferguson include a brief chapter on the stars, for astronomy was then almost wholly preoccupied with the solar system; but in these few remarks of Ferguson's we can find the seeds of many of the ideas William later developed.[57] The season having ended some weeks before, Caroline tells us that he would go to bed with "Smith's *Harmonics* or *Optics*, Ferguson's *Astronomy*, etc., and so went to sleep buried under his favourite authors".[58]

Caroline is writing long after the event, and if William was indeed reading scientific books at the season's end, they must have been by Smith, with the purchase of Ferguson a few weeks in the future. The sequence is what we might expect: as a musician William was of course interested in harmony, and so he may well have thought to buy a copy of *Harmonics* by Professor Robert Smith of Cambridge.[59] If this took his fancy, what more natural than to read another work by the same author, the two substantial volumes of *Opticks*?[60] Smith taught his readers not only the theory of optics, but (and in considerable detail) how to make telescopes, and something of the celestial objects they would then be able to study. A book such as Ferguson's, dedicated entirely to astronomy, would be a logical next purchase.

William's study of *Opticks* was to mark a turning point in the history of astronomy, and in the lives of himself and Caroline. Fortunately, neither Smith nor Ferguson made it clear to their readers that serious astronomers restricted themselves to the study of the solar system, and that the background stars and the universe as a whole should be left to speculators. William therefore saw no reason to discard the plan already forming in his mind, of exploring the structure of the universe. As

Caroline notes with measured understatement, "It soon appeared that my Brother was not contented with knowing what former observers had seen".[61]

For the study of distant (and therefore faint) objects, as William quickly realized, he would need telescopes of great 'light-gathering power'. He first tried refractors, in which the light falls on a lens at the top of the tube and is brought to a focus near the bottom, where it is examined through an eyepiece that is essentially a microscope. On 24 May 1773 William "bought an object glass of 10 feet focal length", the first of many lenses he bought that summer.[62] Caroline records:

> ... he began to contrive a telescope of 18 or 20 feet long.... I was to amuse myself with making the tube of pasteboard against the glasses arrived from London, for at that time no optician had settled at Bath; but when all was finished, no one beside my Brother could get a glimpse of Jupiter or Saturn, for the great length of the telescope could not be kept in a straight line; this difficulty however was by substituting tin tubes soon removed.[63]

However, the limitations of glass technology imposed a severe limit on the size of lenses that could be made. The alternative was to use a reflecting telescope, in which the light passes to the bottom of the tube where there is a mirror to reflect it back to a focus near the top; the image is then (by one means or another) examined through an eye-piece. On 7 June 1773 William notes: "Glasses paid for and the use of a small reflector paid for."[64]

Until the technique of depositing a layer of silver on glass was discovered in the late nineteenth century, mirrors for reflectors were made of alloys known as 'speculum metal', *speculum* being the Latin for 'mirror'. There was no inherent limit to their size: the technology existed for casting large disks which — if he had sufficient skill and patience — a craftsman could hollow out to form a parabolic mirror, and then polish. The grinding and polishing was something William reckoned he could do himself; the problem was, how to get started.

The solution came when he heard of a Quaker living in Bath who had been attempting to polish mirrors for reflectors, but who had now lost interest and wished to dispose of his bits and pieces. On Sunday 22 September William and Caroline took part in the evening servce at the Octagon Chapel as usual; but instead of returning home with Caroline, William went

> to pay him a Visit (by appointment) and there being another gentleman Quaker, their conversation was of nothing else but polishing & polishing-stuffs; and unsuccessful trials. But my Brother bought all their patterns &c though of no further use to him as so many bits of brass or metal.[65]

With these second-hand tools, and with Smith's treatise as his guide, William now set out on the path that would lead to the building of the biggest and best reflecting telescopes the world had yet seen.

William took it for granted that Caroline and Alexander would join in this work. In the early summer, with the departure of William's last students, Caroline had been hoping for a period of peace and recuperation, during which she could get their home into an acceptable state of decoration and cleanliness. No such luck.

> ... to my sorrow I saw almost every room turned into a workshop. A cabinet maker making a tube and stands of all descriptions in a handsome furnished drawing-room. Alex putting up a huge turning machine ... in a bedroom for turning patterns, grinding glasses and turning eye-pieces &c.[66]

Soon William had ordered the casting of his first mirrors. By trial and error he would learn to grind them into shape, and then to impart the polish; "and having advanced considerably in this work it became necessary to think of mounting these mirrors",[67] for without a serviceable mounting the finest reflector would be useless.

Meanwhile music could not entirely neglected, for without it they would starve. In the winter of 1773/74 William decided that his tough little sister needed to acquire some refinement if she was to progress in the Bath musical world. He introduced her to two ladies who were considered good critics of singers, and he coached her in what she should say to them. One was Mrs de Chair, wife of the minister and part-owner of the Octagon Chapel,

> but it was long before I could get into her good graces which was not till I had made her a few of those flattering comp[limen]ts laid in my mouth by my Brother, when she turned round on her heel, saying, Your sister is much improved.[68]

Caroline was distinctly unimpressed. The exact meaning of the comment she would one day insert in her autobiography may not be clear, but the reader gets the message: these people were "the wicked pilfering wretches by which my brother was surrounded".[69]

Undeterred, William thought it would inspire Caroline to greater things if she heard the best singers in the land, and he arranged for her to go to London with one of his pupils, a rich widow named Mrs Colebrook. There were those of his friends who suspected a romance was in the offing;[70] and as it was a rich widow whom William eventually took as his wife, there may have been something in their suspicions. But if so, it came to nothing. Caroline found Mrs Colebrook "very capricious

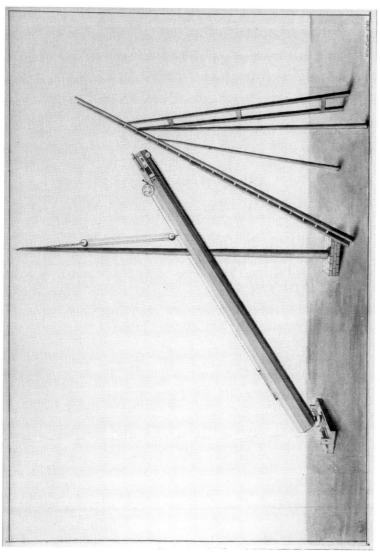

The 'small' 20-ft reflector devised by William in the summer of 1776. The observer stood precariously on the ladder looking sideways through the eyepiece near the top of the tube; if the instrument was near the vertical, the observer would be some 20 feet from ground level, and in the dark. At Datchet, when he had completed the 'large' 20-ft for himself, William set Caroline to work with the small 20-ft, but unsurprisingly she had no success with it. Ink drawing by William Watson made in July 1783, RAS MS W.5/5, no. 4.

and ill-natured",[71] and spent a stressful (and expensive) time in London with her, the planned two weeks extending to six as it proved impossible to return to Bath because of the heavy snow. When at last the road reopened, Caroline occupied the return journey with thoughts of the welcome she would receive from her brothers — only to find herself greeted by "a huge blier-eyed Woman (a new servant)", for Alexander was away and William in bed ill.[72]

By the spring of 1774 astronomy was beginning to challenge music for priority in William's scale of things, and so they moved to a newly built house[73] on Walcot Parade[74] that "afforded more room for workshops and a place on the roof for observing", even though it was "a little way out of town"[75] and so cramped that Alexander eventually moved into lodgings.[76] The landlord, who was a builder and owned timber yards and workshops, lived next door, and William could call on his workmen when required.[77] There was a grass plot behind the house that offered an uninterrupted view to the south, and "preparation immediately was made for erecting a 20 ft telescope",[78] that is, a reflector whose mirror brought the image to a focus 20 feet away, near the top of the tube. The mirror itself was to be 12 inches in diameter, substantial by the standards of the day.

Alexander shared William's enthusiasm. He

> frequently partuck in the laborious and mechanical contrivances and where strength and courage were required he was always foremost to lend his assistance, and even yet I shuder on recollecting the dangerous situation he was in when the gages of the 20ft mirror were struck, standing at the top of the house supporting himself with his left arm on the chimney stack whilst with the right at full stretch he was guiding the plumbline.[79]

Yet Caroline's fond memories of these pioneering days are constantly qualified by regrets at the price she was paying for William's obsession:

> But meanwhile I could not help feeling some uneasiness about my future prospects, for all the time my Brother could spare from his publick business and attending on his scholars was completely filled up with optical and mechanical works; and the fine nights with viewing the heavens, so that I could not hope for receiving any lessons or directions in my practising.[80]

William soon developed skills in making eyepieces of a quality that professional opticians would find beyond belief, but the crude mounting he devised for the 20-ft was unsatisfactory: the telescope was simply slung from a pole, and the observer was perched precariously in the dark on the top of a ladder, peering sideways through the eyepiece.[81] The time would come when William had contrived something better for his own use, and then it was Caroline who would be invited to risk life and limb on this contraption.

Grinding and polishing of mirrors was a craft rather than a science, and when so engaged William did not dare let go and lose his feel for the mirror's shape. Caroline had to feed him:

> ... my time was so much taken up with copying Music and practising, besides attendance on my Brother when polishing, that by way of keeping him alife I was even obliged to feed him by putting the Vitals by bitts into his mouth; — this was once the case when at the finishing of a 7 feet mirror he had not left his hands from it for 16 hours together. And in general he was never unemployed at meals, but always at the same time contriving or making drawings of whatever came into his mind. And generally I was obliged to read to him when at some work which required no thinking; and sometimes lending a hand, I became in time as useful a member of the workshop as a boy might be to his master in the first year of his apprenticeship.[82]

Elsewhere Caroline lists the books she had to read: "Don Quisot, the Arabian night entertainments, Stern's, Fielding's &c novels...."[83]

Alexander meanwhile was not enjoying his life as a bachelor. Caroline and William had already extricated him from one matrimonial engagement, Alexander having caught his fiancée "talking very familiarly with a former suspected lover".[84] Now the builder's pretty daughter was "looked upon as the Bride of Alexander but it was soon found that she had more than one favoured lover, and poor Alex was jilted again".[85] To bury his sorrows Alexander immersed himself in music, to the neglect of all else. His unavailability as a mechanician forced William to curtail his efforts to build telescopes, and at last Caroline had time for singing, albeit entirely on her own: "I applied each moment I could find in practising by myself by given rules; for Musick was seldom thought of."[86]

One musical obligation William could not escape: the requirement to compose a new voluntary for the Sunday morning service at the Octagon Chapel.[87] The choir too was his responsibility, although Caroline helped him at both morning and evening services. Otherwise the highlights of Caroline's week were her shopping expeditions on Wednesday and Saturday mornings, and her dancing lessons.[88]

William's astronomical plans envisaged a still greater part for Caroline to play, and for this she would need more than arithmetic. Her surviving papers include lessons and exercises in the first elements of geometry and algebra. In "A little

Geometry for Lina" the first entry informs her that the angles in a triangle add up to 180°, and that one angle is therefore equal to 180° minus the other two. After seven such 'theorems' we come to "A little Algebra for Lina", which begins by telling her that "$a = b$" means "a is equal to b"; if so, "Equal quantities added to both sides, do not change the equality", and likewise "Equal quantities taken away do not change the equality".[89]

William knew it was important to encourage his pupil: a sheet of exercises of a much later date has been checked by William, who comments, "Bravo Lina! You are right!"[90] However, if we are to believe an anecdote she told in her old age, there were also sanctions: "He used, when making me, a grown woman, acquainted with [mathematical figures], to make me sometimes fall short at dinner if I did not guess the angle right of the piece of pudding I was helping myself to!"[91] In the decades to come, Caroline would have many elementary questions in arithmetic, geometry and astronomy to put to her brother, and she used a commonplace book to write out "the answers ... to the inquiries I used to make when at breakfast, before we separated, each for our dayly tasks".[92]

Thus far Caroline's function in these astronomical activities had been no more than that of a 'gofer', someone who is told "Go for this, go for that!" But before long she had fair warning of the ambitious plans that William harboured for her. An intriguing note in William's "Experiments on the construction of specula" for April 1779 states: "I repolished both speculums of my sister Carolina's Gregorian telescope this morning."[93]

This is the only hint that William was already pressing Caroline herself to become an observer; there is no mention of it elsewhere in either her extensive memoirs of the period or his, and doubtless she then had neither time nor inclination for this additional commitment. William made a similar attempt to involve Alexander, apparently with equal lack of success.[94] This did not stop him one evening from giving a fellow-guest at supper the impression that "His sister and his brother ... were as fond of astronomy as himself and all used to sit up, star-gazing, in the coldest frosty nights".[95]

In the late summer of 1776, by which time the Bulmans had returned to Leeds[96] and Caroline had assumed sole responsibility for the housekeeping, the two men who had first established the Octagon Chapel parted company. Street took control, and appointed a new minister and a new organist.[97] Whether this was with William's agreement or not we do not know — in his memoranda he makes no mention whatever of the termination of his engagement — but the change allowed him to broaden his musical engagements in Bath and Bristol and the surrounding region. By a happy coincidence Linley had recently moved to London, and William took his place as director of the New Assembly Rooms band.[98] But too much of his energy was going into astronomy; although he was an outstanding director of music when his mind was on the job, his concerts were not a success, and a year later Linley would be invited back.[99]

William arranged for performances of the usual Lenten oratorios to be given in the spring of 1777, when the season would be drawing to a close. These could be profitable occasions, for it was common for tickets to be priced as high as five shillings,[100] half a week's wages for a labourer.[101] For such performances many copies of the music would be needed. Isaac Herschel had copied music to support his family in his closing years, William had copied music when he arrived in England as a refugee, and now it was Caroline's turn. She prepared the instrumental parts of *Messiah* and *Judas Maccabaeus* for an orchestra of some seventy or eighty performers,[102] and the vocal parts of *Samson*. She was also responsible for training the trebles in the chorus, and she was herself to sing some of the treble solos.[103] William was concerned that by bending over a desk copying music she might damage her posture as a singer, and so he devised her one at which she could work standing upright.[104] There was some compensation in all this activity: William had less time for astronomy.[105]

As the time drew near for the performances, William was as usual preoccupied with other musical commitments, and so Caroline found herself having to rehearse first the Chapel Boys and then the Choristers as a whole, and this in her own drawing room. "Much confusion and damage" to furniture and fittings resulted, which Caroline would have to make good with the help of a dishonest servant "which I knew tuck every advantage of seeing me so much engaged".[106]

Meanwhile she had to prepare herself to perform as one of the principal singers. Anne Fleming,[107] Bath's preeminent dancing mistress, had been drilling her "for a Gentlewoman" twice a week for a whole year ("God knows how she succeeded");[108] and William gave her ten guineas to buy suitable clothes.[109] On 5 March 1777 Caroline sang as a principal for the first time, in a performance of *Judas Maccabaeus* given in the New Assembly Rooms at Bath.[110] Other oratorios took place on the 12th and the 19th.[111] All went well. The Marchioness of Lothian complimented her "for speaking my words like an English woman",[112] while the proprietor of the Bath Theatre declared her to be an ornament to the stage.[113] Perhaps, against all the odds, she would succeed in making a career as a soprano soloist.

Alas, as soon as the season had ended, William's pent-up enthusiasm for astronomy took over once more, to Caroline's dismay: "of course I saw no other prospect before me but spending the summer in the same manner like the preceding; that is to say — for all the time I could spare from the household business, I found employment in the workroom", or in reading to William while he was at the lathe.[114]

At the end of July 1777 the household was thrown into confusion by word from Hanover that their youngest brother Dietrich had run away from home "with a young idler not older than himself",[115] with the alarming intention of sailing for the East Indies. William at once set off for London, where the son of a Dutch merchant[116] was able to reassure him that time was on his side because no ship would be leaving for the East Indies until late in the year. Suspecting that Dietrich may have returned home on discovering this, he crossed to Holland,[117] and continued to Hanover. There, however, matters took a turn for the worse. "I expect a Letter from Amsterdam every moment", William wrote to Caroline, "and we do already partly believe that he is already engaged to go to the Cape of Good Hope with a very musical gentleman; but nothing is certain yet".[118]

Fortunately the rumour proved unfounded. Dietrich's young companion returned home shortly, with consoling news: he had left his friend in Amsterdam, *en route* for London and Bath. Before long Caroline received a letter to say that Dietrich had fallen sick after arriving in London, and was now at an inn near the Tower. Alexander set off for the capital, supervised the miscreant's convalescence, and brought him by easy stages to Bath.[119] There, on doctor's orders, Caroline fed him on a diet exclusively of roasted apples and barley-water, until William got home and robustly insisted he join them at the dinner table.[120]

When Dietrich was recovered, William found him musical engagements, and he continued to live in Bath with his sister and brothers for the next two years. Caroline had a special place in her heart for her younger brother, but his presence imposed on her yet another distraction from her career as a singer: "... the time necessary for spending on my own improvement was by this addition to the family much broken in upon."[121] For William there were benefits: Dietrich was and would remain a keen entomologist, and he introduced William to the pleasures of moth-collecting.[122] Before another decade passed, William was to scandalize the astronomical world by introducing into astronomy the methods of natural history.

After his return to Hanover, Dietrich married his landlord's younger daughter. Their son Heinrich eventually emigrated to America and died young,[123] but their three daughters remained in Hanover; one day they would antagonize their aged aunt Caroline by what she saw as their neglect of her.

The beginning of the 1777–78 season found Linley in charge once more of the

autumn concerts at the New Rooms, but this time he was facing intense competition from the rival concerts at the Old Rooms, whose managers had enlisted a
talented young Flemish violinist, Franz Lamotte.[124] William meanwhile organized
sixteen private events in aristocratic homes;[125] but again he found no place in
these for Caroline.

The turn of the year brought dramatic changes in the Bath musical scene. Linley
and Lamotte both departed the town, and William found himself in charge of
concerts at both Rooms and in Bristol as well.[126] He enlisted Caroline as a regular
soloist, and for his little sister a career at last beckoned. Curiously, William carefully
notes that in January 1778 he received ten guineas for ten concerts and Dietrich
five, but neither he nor Caroline mentions any payment to her. In March he directed
four concerts of "Airs, Duettoes, and Choruses"[127] and suchlike (for which he and
Dietrich each received four guineas), at the Old and New Rooms alternately. These
were followed, on 15 April, by a performance of *Messiah* for his benefit, with seats
at five shillings each.

Caroline was named as the first soloist, and it seems she gave the performance
of her life, one that could have launched her on a career of her own. Before she had
even left the hall, she was approached with an invitation to sing in Birmingham. In
a fateful response, she declined, "as I never intended to sing anywhere but where
my Brother was the Conductor".[128] The chance had been offered, and spurned,
and would never return.

The next season, 1778–79, Lamotte returned for the autumn concerts, but
William again took charge of those in the spring, and again received a benefit
Messiah. Again Caroline sang; but this time the order of soloists of the previous
year was reversed and she took second place,[129] an early symptom of the decline
of her career. In private William was beginning to cut back on his musical commitments: "I gave up so much of my time to astronomical preparations that I reduced
the number of my scholars so as seldom to attend more than 3 or 4 per day."[130]
Nevertheless, in May he was composing "Glees, Madrigals, Songs and Duettos"[131]
for concerts in the Spring Gardens.

By this time, rumour of the remarkable organist-astronomer of Bath was beginning to spread, as his music pupils passed the word among their friends. As early
as 1774 William had somehow made the acquaintance of Thomas Hornsby, professor of astronomy at Oxford and founder of the Radcliffe Observatory there.[132] In
1777 a neighbour, Dr Lysons, brought no less a visitor than the Astronomer Royal,
Nevil Maskelyne, who was to prove a lifelong friend and ally. He and William had
"several hours' spirited conversation" and only after the departure of the man

'At the NEW ROOMS,

On WEDNESDAY the 15th of *April*, 1778.

WILL BE

Mr. HERSCHEL's Benefit-Concert:

The MUSIC taken from the

SACRED ORATORIO

OF THE

MESSIAH.

The Principal VOCAL PARTS

By Mifs HERSCHEL, Mifs CANTELO, *afterwards Mrs Harris*

Mr. BRETT, Mr. WILSON, & Mr. HERSCHEL.

The FIRST VIOLIN

By Mr. BROOKS, Junior.

☞ *With a Full Chorus of additional Voices and Inftruments.*

N. B. To begin a Quarter before SEVEN o'Clock.

TICKETS at Five Shillings each to be had of Mr. HERSCHEL, No. 19, *New-King-Street*, BATH, and at the Rooms, &c.

Printed by W. GYE, in Weftgate-Buildings, BATH.

Poster advertising the performance of *Messiah* on 15 April 1778, with Caroline as the first principal (and William as fifth). It was after this concert that Caroline was offered, but declined, an engagement in Birmingham that could have marked the beginning of her independent career as a singer. At the corresponding performance of *Messiah* the following year, Caroline took second place to Miss Cantelo. From a private collection.

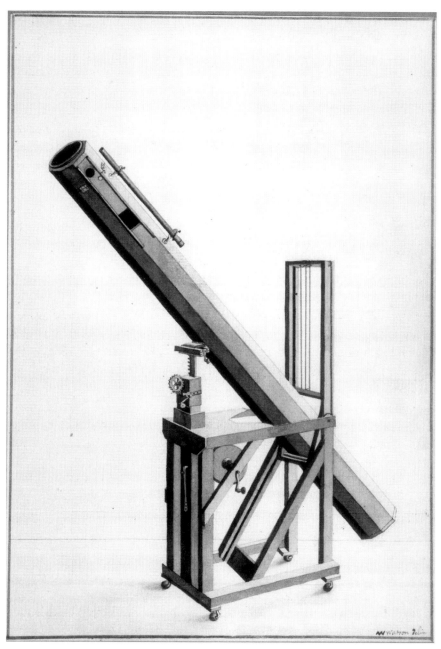

The 7-ft reflector with which William discovered Uranus, from an ink drawing made by William Watson in July 1783. RAS MS W.5/5, no. 3.

whom William described to Caroline as "a devil of a fellow" did she discover the identity of their guest.[133] In late spring of 1778, Edward Pigott, a leading amateur student of variable stars, visited Bath and observed with William.[134] By this time the Herschels had moved back to New King Street. The house in question, no. 19, had a garden that afforded him a good view of the sky to the south, space for the 20-ft, and sheds suitable to be used as workshops.[135]

Early in December 1779 William moved yet again, to Rivers Street, occupying what is now no. 27.[136] Caroline says that this was in order to be near the premises of the recently-founded Bath Philosophical Society,[137] but this must be a mistake as it was only on the 27th that the Society was first mooted.[138] Granted this, the move is difficult to explain. The house had no garden,[139] and so William had to carry his 7-ft reflector into the street when he wanted to observe with it. To erect his 20-ft reflector, for which a permanent site was necessary, he would have to hire a suitable plot of land.

William had by then set himself the challenge of measuring the heights of lunar mountains, and one evening it happened that the Moon was to be seen in front of his house. He therefore carried his 7-ft reflector out into the street and began to observe. A passer-by returning from an entertainment got out of his carriage and asked to be allowed to look through the instrument, "expressed great satisfaction at the view", and embarked on a discussion that went on into the early hours.[140] Next day the stranger called to introduce himself. He was Dr William Watson, Jr, a Fellow of the Royal Society of London, and a member of the infant Bath Philosphical Society.[141] The foundation meeting of the Society took place on the 28th; and it must have been that night, or more likely the 29th or 30th, that Watson encountered William,[142] for William immediately joined the Society and attended the second meeting on 31 December.

He at once found himself among kindred spirits (not least Watson, whose many services to William were to begin with providing a plot of land for the 20-ft[143]); and in the coming months he read his fellow-members no fewer than thirty-one short papers on all manner of topics.[144] The first was on corallines, to which he had probably been introduced by Dietrich. Another reported his "Observations on the Mountains on the Moon", and this latter paper was communicated by Watson to the Royal Society in London. Shorn of William's inappropriate remarks about lunar inhabitants,[145] it appeared in *Philosophical Transactions*.[146] William was beginning to emerge as a figure on the national scientific stage.

For the season 1779–80 William notes laconically, "Concerts at the Rooms as usual".[147] Eastertide oratorios were performed in the theatre at Bath on Wednesdays and repeated in Bristol on Fridays.[148] William notes that "the choruses were under my direction",[149] with the suggestion that the orchestra and soloists were outside his remit; but the poster for the performance of *Alexander's Feast* on 15 March 1780 states, "The Whole under the Direction of Mr. HERSCHEL", and the soloists were supported by "a very full Band and Chorus".[150] The programme also included a Coronation Anthem, a violin concerto, a piano concerto, a madrigal composed by William, and a cello solo played by Alexander: an ambitious evening. One of the soloists came from nearby Wells, but the others were brought to Bath from London or Salisbury, and were no doubt singers of distinction. Standards and expectations had risen since the evening in 1778 when Caroline had been offered the engagement in Birmingham, for then the other soloists had been local, and William — whose talents as a tenor were surely modest — was among them. Meanwhile Caroline's abilities had declined through force of circumstances, and her voice was no longer equal to the challenge of solo performance on such a high-profile occasion. Instead she was restricted to leading the trebles.[151]

William meanwhile was observing with the 7-ft from the garden at New King Street, continuing a programme that he had begun in August 1779 and which was designed to familiarize him with the night sky. He had set himself systematically to examine each of the naked-eye stars and more besides,[152] listing those that proved to be 'double' — pairs of stars so close that at first they look to be single stars. He was having considerable success — for example, on the very first night of the programme he found that the Pole Star was a double star.[153] Although he did not yet realize it, his skills in the grinding and polishing of telescope mirrors, and in the making of eyepieces, were already second-to-none in the world: he was observing with a telescope that would be the envy of professional astronomers when they came to appreciate its excellence. Not only that, but he was becoming experienced in the actual art of observing, and so he was seeing things hidden from all other observers. The result was that when this provincial musician made public his first list of double stars, few would know what to make of his claims, for in other telescopes these same stars appeared unexceptional. One day the President of the Royal Society, no less, would write to congratulate William that an astronomer of unimpeachable authority "has succeeded in verifying your observation of the Pole Star being double".[154]

Early in March 1781 William found Rivers Street so inconvenient that he returned to his previous home at 19 New King Street — but not before he had involved Caroline in an ill-judged business adventure. William had let the ground floor to

19, New King Street, Bath, the home of William, Caroline and Alexander from September 1777 to December 1779 and again from March 1781 (when William discovered Uranus) to July 1782. It was in the basement of this house that William and Alexander attempted to cast the disk for a 3-ft mirror in August 1781. The building is now the William Herschel Museum. Photograph by Michael Tabb, 1984.

people with a millinery business, in which he bought Caroline a share. She was to keep the books and make sure "nothing went wrong"; but the location was unsuitable, dishonesty was rife, and the business failed. The loss to the Herschels, she says, was trifling, but she had to remain on the premises until the stock had been sold off.[155]

On 13 March, while Caroline was still at Rivers Street, William's programme took him to the stars near Zeta Tauri, and he immediately noticed one of unusual appearance. He suspected it might not be a star at all, but an unknown member of the solar system, and so presumably a comet. Re-examining it four nights later, he found that his suspicions had been well-founded, for the object had altered its position in the sky; no distant star could possibly move so far in so short a time.[156]

Watson helped him to share the discovery with the wider scientific community.[157] On hearing of the supposed comet, Maskelyne and Hornsby did their best to locate it with the help of William's inexpert description of its position; but frustration followed, for only in William's home-made telescope could the object be distinguished from a normal star. Maskelyne was reduced to studying the stars (real or supposed) in the relevant region of sky, and going back later to see which of them had moved.[158] When finally he identified William's object, it did not seem to him to have the appearance of a comet; indeed, he thought it might even be a planet.

Months later, when there were enough observations of the object for mathematicians to compute its orbit, Maskelyne would be proved right. It was indeed a planet, the first to be discovered since the dawn of history. William was catapulted overnight into national and international fame, his hitherto-unknown name a puzzle to editors all over Europe. He was variously Mersthel,[159] Hertsthel,[160] Hertschel,[161] Horochelle.[162]

Meanwhile William had a living to earn. On 21 March, just four days after he had confirmed that his object was indeed a member of the solar system, he took part in an oratorio performed in Bath (and repeated in Bristol two days later).[163] The 21st was also the day when Caroline completed the winding-up of the millinery business and at last was free to rejoin her brother at New King Street; but she makes no mention of these oratorios and presumably had no role in them.

However, Easter was just over three weeks away, and the oratorios that were to be given in Passion week[164] — two in Bath and two in Bristol[165] — called for careful rehearsal. Caroline was to be responsible once again for training the sixteen or so trebles, but her days as a principal singer were past. She had performed as a soloist for the first time in 1777; in 1778 she had been leading soloist, and had been offered (but declined) an engagement in Birmingham; in 1779 she had been second soloist. But in 1780, and again in 1781, she was no more than trainer and leader of the trebles (sopranos).

> The solo performers he had engaged from London, and I only lead the treble as I did the season before; for the interruption in my practice of the preceding months, besides accumulation of copying music &c. left me no time to take care of myself [as a singer] or to stand upon nicety's.[166]

Once the season ended, William's passion for astronomy took over their lives completely:

> ... my Brother applied himself to perfect his mirrors erecting in his garden a stand for his 20-ft. Telescope; which caused (I remember) many trials to be made before the required motions for such an unwieldy machine could be contrived; of which many (I suppose) were made by way of experiment against a mirror for an intended 30-ft. Teles. should be completed, for which between whiles (not interrupting the observations with 7, 10, and 20 feet, and writing papers for both the Royal and Bath Ph. Society's), Gages, Shape, weight &c. of the mirror were calculated, and trials of the composition of the metal were made. In short I saw nothing else and heard nothing else talk but about those things when my Brothers were together. Alexander was always very alert and assisting when anything new was going forward, but he wanted perseverance and never liked to confine himself at home for many hours together; and so it happened that my Brother was obliged to make trial of my abilities in copying for him Catalogues, Tables, &c. and sometimes whole papers which were lent him for his perusal ... which kept me employed while my Brother was at the telescope at night; for when I found that a hand sometimes was wanted when any particular measures were to be made with the Lamp micrometer &c. and a fire kept in, and a dish of Coffe necessary during a long nights watching; I undertook with pleasure what others might have thought a hardship.[167]

At first William's ambitions for the proposed monster reflector called for mirrors no less than 4-ft in diameter. This he reduced to 3-ft, but even so the mirrors were to be the largest in the world. No local foundry could even cast such a disk. In no way daunted, and disregarding the risk of a general conflagration, William set about converting the basement of the house in New King Street into a foundry. "The mirror", Caroline writes, "was to be cast in a mould of loam prepared from horse dung of which an immense quantity was to be pounded in a morter and sifted through a fine seaf; it was an endless piece of work and served me for many hours exercise and Alex frequently took his turn, for we were all eager to do something towards the great undertaking".[168] Even William Watson, M.D., F.R.S., future knight of the realm, found himself pounding horse dung.[169]

Caroline was excused from taking part in the dangerous work of casting. The first of the two attempts that William made in August 1781 proved fruitless when the mirror cracked on cooling.[170] During the second, the mould broke and the molten metal poured onto the flagstones. Caroline, viewing proceedings from the safety of the garden, records that "both my Brothers, and the caster and his men were obliged to run out at opposite doors, for the stone flooring (which ought to have been taken up) flew about in all directions as high as the ceiling".[171] William had barely escaped with his life, and now even he was prepared to postpone further attempts.

That November he received the Copley Medal of the Royal Society for his discovery of a "new star", be it planet or comet.[172] The award was to be conferred during the height of the season at Bath, and William could attend the ceremony only by taking the night coach to London and returning as quickly as possible.[173] Within days he was elected Fellow of the Society.[174] In Sir Joseph Banks, who was President, and Nevil Maskelyne, the Astronomer Royal, he had admirers with influence at Court, and the sensational discovery gave them all the leverage they needed.[175] The King had in fact known of William's discovery since the previous summer, for back in August he had personally mentioned it to Dr S. C. T. Demainbray, who had been the King's tutor long ago and was now astronomer at the King's private observatory at Kew; and Demainbray had written to William to say that he was welcome to come to Kew and use the instruments there to measure the position of his "Comet or Planet".[176] But William's allies were probably unaware that the King and Queen had long been familiar with the Herschel family, for, as we have seen, the Queen had conferred patronage on Jacob Herschel back in 1769.[177] Sophia's eldest son George Griesbach arrived in London in 1778 to become a member of the Court orchestra, and one evening he played a concerto of Jacob's. The King enquired who had given it to him.

> "My uncle." "Who is your uncle?" "Herschel, at Hanover." Here the King left me and went to tell the Queen whose nephew I was.[178]

As Jacob had well understood, all over Europe the relation between client and patron had long been subject to convention: the client dedicated a work to the patron, and received tangible advantage in return. Thus, in 1610, Galileo had named the moons of Jupiter the 'Medicean stars', and had become Mathematician to the Grand Duke. Banks thought that William would do well to imitate Galileo, and that his new planet should be "sacrificed somehow to the King", by means of a dedication.[179] If the King accepted the dedication, custom required him to confer a benefit in return — though it was not easy to see what form this benefit should take.

Banks's first thought was that William might succeed the elderly Demainbray at the "snug" private observatory at Kew.[180] The position would indeed have been ideal for William, but Demainbray unwittingly threw a wrench in the works. As Banks told Watson on 23 February 1782, "as the Devil will have it he died last night". The dedication would now come too late. "I fear [the time] has passed by which a well timed compliment might have helped if the old gentleman had chose to live long enough to have allowed us to have paid it."[181] In the hope that appointment to Kew might still be possible on merit alone, Banks attended the royal Levy the following morning, "but did not receive any hopes" — the reason being, it later transpired, that the King had already promised the succession to Demainbray's son.[182]

By now the campaign by William's allies was in full swing, and he was "informed by several gentlemen that the King expected to see me".[183] Not only that, but he was to bring his telescope. But the season was not yet ended, and he had musical commitments he could not escape. In Passion week four oratorios were given, two in Bath and two in Bristol, with William as director and chorusmaster; but it seems that Caroline again played no part.[184] William had taken financial responsibility for the series, and according to Caroline he lost heavily.[185] Not surprisingly, he was distracted by events in London, and the organization was chaotic. On 7 March the *Bath Chronicle and Weekly Gazette* announced that the Wednesday oratorio in Bath would be *Jephtha*. On 14 March this was amended to *Samson*, and on the 21st, less than a week before the performance, to *Judas Maccabaeus*.[186]

The Bristol performance on the Thursday was nothing short of a disaster. In the morning William was deep in conversation with Watson about the royal summons when his eldest nephew George Griesbach arrived from Windsor with "confirmation that his Oncle was expected with his Instrument in Town".[187] A chaise was at the door ready to take them to Bristol for the morning rehearsal, for which they would need the musical parts for nearly a hundred performers, but William had thoughts only for London, and Caroline had to assemble the music as best she could. A letter to the manager of the Bristol theatre, published in *Bonner and Middleton's Bristol Journal* on Holy Saturday, describes what ensued that evening:

Perhaps no audience was ever more impos'd on, or worse treated than that which Thursday night attended the performance of the *Messiah* at your theatre.

Many Gentlemen who went principally with a wish of hearing Mr. Tenducci, found themselves at the drawing up of the curtain (and not till then) disappointed — Hand-bills indeed were printed; but they were confin'd wholly to the company of the boxes — and so were only printed with a view to save the Manager's credit — It was expected and hop'd that some exertions would have been made by Mr. Rauzzini to compensate for Mr. Tenducci's absence — but that performer satisfied himself with singing one song, and joining, now and then, in a chorus. — Such was Mr. Herschel's eagerness to conclude the performance, that songs — duets — choruses, were omitted — the audience disgusted — and the band thrown into confusion. The first violin led off one air, while the violoncello had begun the accompaniment of another.

The chorus singers were repeatedly at a loss whether to stand up or keep their seats; and Mr. Rauzzini had almost trampled Miss Storer to death, in endeavouring to sing from Mr. Croft's paper, instead of his own, which neither himself or the conductor of the band knew anything of.[188]

The following week William responded with an apology for the change of singer and the delay in distributing handbills; but he made no comment on the other criticisms.

On 1 May a chastened William directed another performance of *Messiah*, this time to mark the inauguration of a new organ at St James's, the largest parish church in Bath. Lessons had been learnt from the Bristol *débâcle*, and this time the choir was strongly reinforced, by choristers from Salisbury along with the famous singers of Lady Huntingdon's Chapel; and the performance was a success.[189] Again there was no place for Caroline,[190] but on 19 May, Whitsunday, when one of William's anthems was performed in St Margaret's Chapel at Bath with the composer at the organ, Caroline sang the treble solo.[191]

This was to be the last time either of them performed in public. Early the following week[192] William set off to meet his mentor Watson at the London house of Watson's father, where he was to stay. The 7-ft reflector with which he had discovered the planet had not been designed to be portable by coach, so William had had to devise a new stand and steps so that the instrument could be disassembled and carried in a box. He also took an atlas, his recently-published catalogue of double stars, and everything else he might need to impress the King.[193] On the Friday William had dinner with a distinguished company that included the Astronomer

Royal, Alexander Aubert (a leading amateur observer and the one who had verified that the Pole Star was double), and Colonel John Walsh of the Worcester Regiment.[194] Walsh it was who had pointed out to the King on 30 April that William had a double claim for permission to dedicate the planet to the ruler of both Hanover and Great Britain, since he was a Hanoverian resident in Great Britain.[195]

George III had been on the throne since 1760, and he was to rule until 1811 when his "madness" forced him to yield power as Regent to his son, the future George IV. George III was a shrewd man who believed in hands-on government, and he personally interviewed William on the Saturday. The discovery of the planet was prima facie proof of the excellence of William's reflector; but the King wanted confirmation from his Astronomer Royal and other leading observers, and the way to do this was to have the reflector set up at Greenwich alongside the instruments of the Royal Observatory, for a trial of excellence. After that the King and his family would enjoy taking a look through it themselves.

> Dear Lina, I have had an audience of His Majesty this morning and met with a very gracious reception. I presented him with a drawing of the solar system, and had the honour of explaining it to him and the Queen. My telescope is in three weeks time to go to Richmond [near Kew], and meanwhile to be put up at Greenwich, where I shall accordingly carry it today.[196]

Caroline is to tell his music students that he cannot resume lessons until he has the King's permission to return. His absence is the likely reason for the chaos in the arrangements for the annual Pauper Charity concert in Bath; the advertisement that appeared the day before the performance was still unable to give details of the programme to be performed.[197]

On Wednesday, 29 May, William assembled his reflector at Greenwich.[198] On the Friday the King invited him to his regular concert, chatted with William for half an hour, and asked George Griesbach to play them a solo concerto.[199] Then it was time for the trial.

> These two last nights I have been star-gazing at Greenwich with Dr Maskelyne and Mr Aubert. We have compared our telescopes together, and mine was found very superior to any of the Royal Observatory. Double stars which they could not see with their instruments I had the pleasure to show them very plainly.... Among opticians and astronomers nothing now is talked of but *what they call* my Great discoveries. Alas! this shows how far they are behind, when such trifles as I have seen and done are called *great*. Let me but get at it again! I will make such telescopes, and see such things....[200]

His London visit was proving pleasant and satisfying in its way, but William was already in his middle forties and he had so much to do and so little time in which to do it. King George was charming to a fault, but time was passing.

The King was in something of a quandary. The Kew post was filled, so what was he to offer William? At some stage the suggestion was floated of a post in Hanover, where there was as yet no royal astronomer. But the salary mentioned was a mere £100 per annum, barely a quarter of what William had been earning from music for years. Such a suggestion could hardly be taken seriously, and indeed we know of the proposal only from a letter that Caroline wrote sixty years later.[201]

The way out of the difficulty became clear when William was at last summoned to Windsor Castle to demonstrate his telescope to the King and Queen and a host of other members of the Royal Family. The viewing extended over at least three nights, beginning on Tuesday 2 July, when his reflector was set up alongside two refractors by Dollond and a reflector by Short,[202] leading makers of the day. "My Instrument gave a general satisfaction; the King has very good eyes and enjoys Observations with the Telescopes accordingly."[203]

William was skilful in personal relations. Thus on the second night, when the King and Queen were absent and the princesses asked what they might see without going out on the grass, William took his telescope indoors and together they waited for Jupiter or Saturn to appear. When clouds interfered with their viewing, William produced a pasteboard Saturn that he had prepared beforehand, and "The effect was fine and so natural that the best astronomer might have been deceived".[204]

Complex negotiations now took place 'off the record', for protocol demanded that no offer from the King should be refused, and therefore His Majesty had to know in advance that the offer he was about to make was acceptable.[205] The out-come was that William was awarded a pension (that is, a salary for life[206]) of £200 per annum, his only obligation being to live near Windsor, available to show the Royal Family the heavens when they so wished. Otherwise he was completely free to devote himself to astronomical research.[207]

William's dedication[208] of the planet to the King would now be by way of 'Thank you', rather than the usual 'Please' of patronage convention. To him it would always be Georgium Sidus; but in time the astronomical community came to prefer the more conventional name of Uranus.

No more dreary music lessons for pupils of little or no talent, no more training of indifferent choirs; instead, a sinecure that would allow him to devote himself to astronomy for the rest of his days. It was an offer William could not have refused. True, £200 was much less than he was currently earning from music — Watson commented, "Never bought Monarch honour so cheap!"[209] — but it was equal to two-thirds the salary of the Astronomer Royal and *he* had to work hard for his money.

Besides, William knew that Maskelyne had already commissioned a cabinet-maker to produce a copy of the stand of William's 7-ft.[210] This was something that William himself could have taken responsibility for. But why stop there? Could not William supplement his pension by making complete telescopes for sale? He would have no need to advertise: the planet and his double stars were incontrovertible testimony to the quality of his instruments. The very next month, Christian Mayer, court astronomer at Mannheim and himself a student of double stars, would write to ask, "Could you send me, at a price fixed by yourself, one of your telescopes?"[211] And in October 1783 the King himself gave his blessing to this profitable sideline, by placing an order for no fewer than five of William's 10-ft reflectors[212] (although the blessing proved mixed, for as late as February 1791 the King had still not paid for any of them[213]). William looked back with satisfaction on the origins of his career as a commercial telescope maker:

> The goodness of my telescopes being generally known I was desired by the King to get some made for those who wished to have them. Getting the wood-work done by his Majesty's cabinet maker, I fitted up five 10 feet telescopes for the King, and very soon found a great demand for 7 feet reflectors. This business, in the end, not only proved very lucrative but also enabled me to make expensive experiments for polishing mirrors by machinery.[214]

So William had accepted, and gladly. Since he was already at Windsor, he scouted around for a suitable home, and found one that met his requirements in the village of Datchet, within sight of the Castle. In less than a fortnight he and Caroline were installed there, their careers in music a thing of the past.

It was William's own decision, and indeed the fulfilment of his ambition. The performance of his anthem on Whitsunday had been when "for the last time I performed on the Organ",[215] but he had no regrets whatever. When he returned to Bath to arrange the move, he set about extricating himself from his musical commitments with enthusiasm, for "the prospect of entering again on the toils of teaching &c. which awaited him at home ... appeared to him an intolerable waste of time".[216]

Caroline similarly notes that in singing the treble part on Whitsunday, "I opened my mouth for the last time before the public".[217] William did not explicitly consult her before accepting the King's offer, but they had both known the purpose of his summons to London, and she must have assured him in advance of her support. But the enormity of the decision she had made finally struck her when she had been at Datchet for a month, and Alexander, who had helped in the move, returned to Bath for the beginning of a season in which she would have no part. It was not music itself that was the real sacrifice, but the possibility of earning her own living. In Hanover, between the death of her father and her rescue by William, Caroline had known what it was to be financially dependent, on a brother who did not hesitate to hand out whippings and who would doubtless have no further use for her after he was married. The ability to earn her own living, be it as governess, singer, harpsichordist or whatever, had long been an obsession. When Alexander left for Bath,

> The separation was truly painful to us all, and I was particularly affected by it; for till now I had not had time to consider the consequence of giving up the prospect of making myself independent, by becoming (with a little more uninterrupted application) a useful member to the musical profession. But besides that my Brother would have been very much at a loss for my assistance, I had not impudence enough for throwing myself on the public after losing his protection.[218]

Helping William in his new vocation offered her a further opportunity to repay her debt to the bachelor brother who had rescued her from servitude, and over whose home she presided.

It seems it did not occur to her that William might not remain a bachelor for ever.

3

"A Female Astronomer"

Moving house is never easy, and it took several days of effort before the contents of 19 New King Street were loaded and ready to begin their journey to Datchet in the early hours of 31 July 1782. William followed on the 11 a.m. coach to London, and alighted at Slough, then a tiny village north of Windsor. There he would no doubt have stayed the night at the Crown Inn, whose owner he would one day be.

Next day he was joined by Caroline and Alexander, and they all took a meal at the Inn before setting off on foot for Datchet, along the Windsor Road. After a couple of minutes they passed a house on the left where William was to spend the last 36 years of his life.

On arrival at Datchet they spent the night at the Five Bells inn near the church.[1] When they awoke they found to their relief that the wagon with their effects and William's astronomical equipment had safely arrived, and so they walked over to see their furniture unloaded and installed.

Caroline was appalled at the sight of her new home, "the ruins of a place".[2] She tells us that it had once been a gentleman's hunting seat, though this is puzzling in view of its location. It was in fact an annex to what is now "The Lawns" in Horton Road, then a rambling property but since reduced and remodelled. A newspaper advertisement inserted when William eventually left tells us that the annex consisted

> on the first floor, of four convenient bed-chambers, with garrets; on the ground floor, an hall, two parlours, kitchen, larder, beer cellar, laundry, and wash-house; coach-house, and stables, with a garden walled in.[3]

But when the Herschels arrived it had not been lived in for years. For the next two months they would be surrounded by workmen trying to make the place habitable. The garden was deep in weeds, and when Alexander looked around for a suitable site for the 20-ft telescope he almost fell down a well. They had brought no servant with them because William had hired one locally, but of her there was

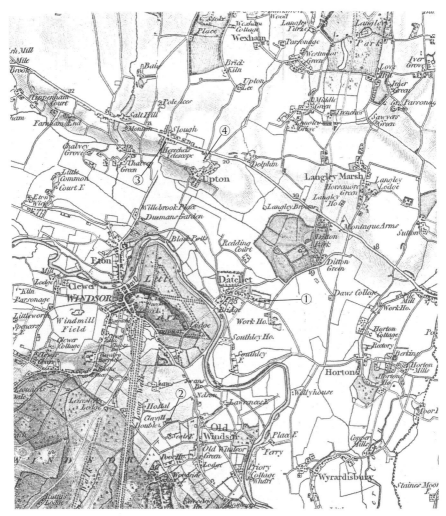

The first Ordnance Survey map of the Windsor area, published in 1830. (1) William and Caroline's house at Datchet, (2) Clay Hall at Old Windsor, (3) their home at Slough, (4) the Upton home of John and Mary Pitt. To the east of (4) is the Dolphin, one of the properties in Langley Marsh owned by John Pitt.

no sign. Enquiries revealed that she was in prison for theft. Prices in the shops were horrendous,[4] the butcher dishonest,[5] coal twice the price charged in Bath, taxes on a building with thirty windows crippling; and they had only £200 a year coming in.[6]

William and Alexander did not see it this way. The stables would do for grinding mirrors, the laundry would make a good library, there was everything a telescope maker and astronomer could wish for. William could not wait to get started. They arrived on the 2nd, and by the 5th he had resumed observations; after all, he was now a professional astronomer.

Unfortunately, the Royal Family had so much enjoyed looking through William's telescope that he found himself regularly summoned to the Castle. This was part of the deal; but it tried his patience (and the depth of his pocket), because he would then have to pay to transport his 7-ft home if he was to use the rest of the night to add to his list of double stars.

A pause was agreed while William returned to Bath to collect the fragments of the 3-ft mirror and other equipment, and this served to break the habit. Summonses to attend the Castle became fewer. In part this was because the novelty of having an astronomer on call began to wear off, but in part because the Royal Family realized that the portable 7-ft was a small telescope by William's standards, and that to take full advantage of the presence near Windsor of the King's astronomer it was they — and their innumerable guests — who must visit him.[7] Four months after Caroline first set foot in Datchet, she was hostess in her delapidated home to the King himself, come to look through William's 20-ft.[8] The Royal Family tried to be considerate about their visits — the Queen on one occasion postponed an appointment from Saturday to Sunday, but only "provided you do not think it a sin to look at the planets on a Sunday's evening".[9] William and Caroline became accustomed to welcoming their sovereign and his retinue to their home, and a visit to Herschel became a routine solution to the problem of how to amuse royal guests. But the King was serious about astronomy, and sometimes would call with only an equerry or two for company.[10]

Among the first objects William came across when he began observing at Datchet was a "nebulous star or telescopic comet".[11] This proved to be one of the hundred or so milky patches or 'nebulae' that had been listed recently by the French comet-hunter Charles Messier, Astronomer of the Navy of France, so that he would not confuse these permanent features of the night sky with a newly-arriving comet.[12] So obsessed was Messier with comets that he lamented missing one when he had to nurse his wife on her deathbed.[13]

William had seen just four nebulae all the time he was in Bath, but he had long been intrigued as to their nature.[14] On 30 August he showed two of them to the King.[15] Clearly a cluster of stars, so far away that telescopes could not distinguish the individual members of the cluster, would appear as a milky patch, so he could be confident that some nebulae were simply star clusters. But were there nebulae that were not star clusters, composed instead of some sort of luminous fluid? This was a question he must try and answer.

Caroline meanwhile was doing her best to make herself useful:

> I had ... the comfort to see that my brother was satisfied with my endeavours to assist him when he wanted another person, either to run to the clocks, write down a memorandum, fetch and carry instruments, or measure the ground with poles, &c., &c., of which something of the kind every moment would occur.[16]

It then occurred to William that Caroline, "my assistant in astronomy",[17] would herself become familiar with the heavens, were she systematically to sweep the night sky on the lookout for interesting objects; and it would be a useful way for her to occupy the hours of darkness.[18] Typically, he did not ask if this would appeal to her, he simply told her that this was what she was to do.

She was not best pleased. Her career was in ruins; and in place of her busy and sociable life in the town of Bath, where there were pupils arriving throughout the day whom she had to welcome, choirs assembling for her to train, and performances for which she had to practise, she was now leading a lonely, hermit-like existence in a tiny village. When William was out of the house she was

> left solely to confuse myself with my own thoughts, which were anything else but cheerful; for I found I was to be trained for an assistant Astronomer and by way of encouragement a Telescope adapted for sweeping consisting of a Tube with two glasses such as are commonly used in a finder [was given to me], I was to sweep for Comets, and by my Journal N° 1, I see that I began Augt 22, 1782 to write down and describe all remarkable appearances I saw in my Sweeps (which were horizontal). But it was not till the last two months of the same year before I felt the least encouragement for spending the starlight nights on a grass-plot covered by dew or hoar frost without a human being near enough to be within call.[19]

Caroline's recollection in old age was that William instructed her to sweep for comets, and in time this was to become her primary goal. But the original list of

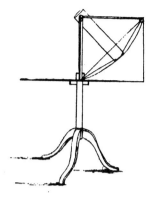 

Sketches by W. H. Smyth, 1843/4, of Caroline's first two sweepers. (*left*) The "Tube with two glasses" adapted for horizontal sweeps, 1782; (*right*) the sweeper William made her in 1783 for vertical sweeps: by turning the handle and so winding in the string (which goes vertically up, then horizontally along the top of the panel, and finally down to the bottom of the tube), Caroline was able to raise the tube from the zenith to the horizon while keeping her eye to the eyepiece. MS Gunther 36, ff. 124v, 125v, Museum of the History of Science, Oxford University.

objects for which William wished her to search was much more general:

> To be wrote down.
>
> Double stars that appear to be one, two, or three diameters asunder.
>
> Clusters of stars such as 5, 6, 7, 8, &c. near together, all within a dozen diameters or so.
>
> Nebulas.
>
> Comets.
>
> In setting down such Phenomena they must be described by lines from certain stars and figures drawn upon paper — for example, "I see a nebula, its situation is pointed out by a line drawn from *A* to *B* crossed by another line from *C* to *D*."[20]

And indeed the only interesting object she found on her first night was one of William's double stars.[21]

Her second night, 30 September, netted her several double stars together with a nebula.[22] This proved to be the 27th in Messier's list (M 27), but it was one that William had not yet seen: Caroline had something interesting to show her brother. On 2 October she succeeded in recording the position of one double star, but she ran into problems with others: "Saw a great many double stars; but the telescope filled so fast with dew, that I could not find time to mark their places."[23]

On the 4th and the 12th she swept again, but her efforts yielded little more than

23 Canis majoris. My Brother examined it with 460. and found not less than 20 stars. with 227 above 40. with a compound eye piece perhaps 100 or 150 and very beautiful, nothing nebulous among them. Messier has it not.

Following γ Canis majoris a very faint Nebula, my Brothers observation upon this nebula; About 3½ degrees following γ Canis maj Amas d'étoiles. It is about ½ deg following a star of the 7th or 8 magnitude. with 460 there are about 15 or 16 stars. which are all excessively obscure, and seem a little nebulous; but I think it owing to the low situation and high power. with 227 about 40 or 50 small stars. with the compound piece, a cluster of small stars closer than those in the foregoing Nebul Messier has it not.

Observed the nebula near the 2 Navis. a cluster of bright stars. (Mess. 46.)

Saw the nebula near Canis majoris, the stars are not so numerous and bright as in the 46th (Mess. 41st)

Caroline's notes of her sweeps on 26 February 1783. Of the four nebulae she came across, two (M 46 and M 41) were in Messier's list, but of two others she can say "Messier has it not". In those few minutes Caroline had increased the number of nebulae known to astronomy by two percent. RAS MS C.1/1.1, 5.

three double stars near Gamma Pegasi.[24] On the 13th, however, she came across another nebula, which proved to be M 36, and on the 29th, M 13 and M 37.[25] As winter advanced her evenings of sweeping became very occasional, but her first entry of the new year, 23 January 1783, betrays more than a dutiful interest:

> Look'd at the Nebula in Cancer [Praesepe, the Beehive Cluster] for half an hour, saw nothing nebulous; but they are all very distinct little stars. Began at the above Cluster and made a horizontal sweep all around, I spent above an hour upon this sweep, but found nothing remarkable. It was too cold, therefore could not continue my observations. A pretty strong frost.[26]

Despite the weather, Caroline was beginning to enjoy sweeping.

26 February proved a turning point. "Nebula, about 1¼ deg. north preceding the bright star in the Ship preceding the 1st Navis of Fl[amsteed] towards 23 Canis Majoris. My Brother examined it.... Messier has it not."[27] A few minutes later she continues: "Following γ Canis Majoris a very faint nebula.... Messier has it not." She then came across what proved to be M 46, and she rounded off the evening with M 41. Her next night, 4 March, yielded two more nebulae unknown to Messier;[28] and the night after that, 9 March, a cluster in Monoceros. Within months of her last performance as a soprano, and armed with a telescope that was little more than a toy, Caroline had increased the number of nebulae known to astronomy by five percent. The modern reader turning the manuscript pages of her Books of Observations can scarce forbear to cheer.

Caroline was even planning a catalogue.[29] Mightily impressed, William replaced her primitive refractor with an instrument he had designed and built specially for her to use in sweeping, and it was small enough for her to use sitting down.[30] It was

> ... a Newtonian reflector, of 2 feet focal length; and with an aperture of 4.2 inches, has only a magnifying power of 24, and a field of view 2°12′. Its distinctness is so perfect, that it will shew letters at a moderate distance, with a magnifying power of 2000; and its movements are so convenient, that the eye remains at rest while the instrument makes a sweep from the horizon to the zenith.[31]

In 1791 William (by then an honorary Doctor of Edinburgh University[32]) was to present her with a more powerful sweeper of broadly similar construction, and a 1793 letter from Nevil Maskelyne gives us an excellent account of how she worked:

> I paid Dr & Miss Herschel a visit 7 weeks ago. She shewed me her 5 feet

Newtonian telescope made for her by her brother for sweeping the heavens. It has an aperture of 9 inches, but magnifies only from 25 to 30 times, & takes in a field of 1°49′ being designed to shew objects very bright, for the better discovering any new visitor to our system, that is Comets, or any undiscovered nebulae. It is a very powerful instrument, & shews objects very well. It is mounted upon an upright axis, or spindle, and turns round by only pushing or pulling the telescope; it is moved easily in altitude by strings in the manner Newtonian telescopes have been used formerly. The height of the eye-glass is altered but little in sweeping from the horizon to the zenith. This she does and down again in 6 or 8 minutes, & then moves the telescope a little forward in azimuth, & sweeps another portion of the heavens in like manner. She will thus sweep a quarter of the heavens in one night. The Dr has given her written instructions how to proceed, and she knows all the nebulae [listed by Messier] at sight, which he esteems necessary to distinguish new Comets that may appear from them. Thus you see, wherever she sweeps in fine weather nothing can escape her.[33]

The warmth of the admiration of the Astronomer Royal is evident, as is the ingenuity of William's design and Caroline's dedication in using the instrument. Yet we note that even after a decade of this exceedingly simple procedure, Caroline still needed the crutch of her brother's written instructions.

William had rightly advised Caroline that as a comet-hunter she must familiarize herself with the Messier objects. In her early weeks of observing she would often spend the bulk of her time in sweeping, and then round off proceedings by looking for one or two of Messier's nebulae. Her entry for 6 August 1783 is typical:

I should have begun to sweep with δ Aquilae & β Cygni; but as the evening was too far advanced, I began with β Cygni. I left off with μ Aquarii. The night being indifferent I look'd for some of Mess. Nebulae, and saw 81 & 82; but could not see 97; nor that near δ Ursae Majoris. Cloudy.[34]

By the end of the year she had seen about a third of them. It was while examining the Andromeda Nebula, M 31, on 27 August that she made the notable discovery of its companion now known as NGC 205.[35]

Caroline's early encounters with nebulae had opened William's eyes to the harvest that awaited the two of them, and on 4 March he broke off his self-imposed task of examining every star in John Flamsteed's great catalogue and himself "began to sweep the heaven for nebulas and clusters of stars".[36] Instead of one of his own reflectors he used a small refractor he had purchased; but before long the absurdity of using this modest instrument to find nebulae struck him. A comet-hunter had no time to lose, and must try to scan the whole of the visible heavens in as few nights as possible; for this a telescope of moderate size was appropriate. But the nebulae were permanent features of the night sky, and so there was no rush to detect them. It was more important to take one's time and subject each nebula to detailed scrutiny with a major telescope — such as the one he was now building.

When he moved to Datchet, his reflector with the greatest 'light gathering power' was still the 20-ft with its 12-inch mirrors; but its mounting was primitive and the observer was perched on a ladder in constant danger of falling. The telescope was wholly unsuitable for a sustained campaign. William accordingly had turned his inventive mind to the task of contriving a secure and stable mounting; and a year after their arrival at Datchet he completed one of the great telescopes of all time. It was a 20-ft like its predecessor, though this time with mirrors 18-inches in diameter. More importantly, it had a wooden mounting on wheels, which could be rotated (by a single workman) to face in any direction; and this incorporated a seat for the observer that could be raised and lowered on ladderwork installed for the purpose. Later the seat would be replaced by a platform that extended from one side of the tube to the other.[37] The mounting was a triumph of ingenuity, and the resulting instrument was ideal for the systematic search for nebulae that William planned, a search that would extend over the entire sky visible from Windsor.

In August 1783 the 'large' 20-ft, as it came to be known, was nearing completion, and William could not wait to get started. Perseverance he had in abundance, patience almost none.

> My Brother began his series of Sweeps when the Instrument was yet in a very unfinished state, and my feelings were not very comfortable when every moment I was alarmed by a crack or fall, knowing him to be elevated 15 or 16 feet on a temporary cross-beam instead of a safe gallery. The ladders had not even their braces at the bottom and one night in a very high wind he hardly had touched the ground before the whole apparatus came down.[38]

William's completion of the large 20-ft had now released the small 20-ft for Caroline's use, and for all its failings this was a major telescope. William wondered what she might best do with it? Comet-sweeping was a hit-or-miss affair; the nebulae he had taken over himself; but double stars offered an *entrée* into two of the great unsolved problems in stellar astronomy: the distances of the stars, and whether Newtonian attraction operated beyond the solar system. If Caroline re-examined his doubles, there was no knowing what might emerge.

A watercolour (in private possession) showing the 'large' 20-ft at Datchet. The instrument could be rotated on wheels, but in use it was usually directed to the south. Structural ladderwork is seen to what would then be the east and the west, but on the east side additional ladderwork supports an observing chair that can be raised and lowered.

The problem of how to measure the distances of the stars had long been in William's mind, and it had been a primary motivation in his own search for double stars. By the mid-eighteenth century the general order of magnitude of stellar distances was well understood. Newton and others had achieved this by supposing that the stars were physically uniform, and that the Sun looked brighter than Sirius only because it was nearer. If this assumption was correct, then to find out the distance of Sirius, one need only ask how far away the Sun would have to be in order to appear of the same brightness as Sirius. Newton reckoned the answer to be some one million times the present distance of the Sun.

But this was no more than an estimate based on a questionable assumption. To make actual measurements, astronomers would have to exploit the fact that the Earth's journey around the Sun takes us every six months from one side of the Sun to the other. If therefore we measure the apparent position of a star in January, and measure it again in July, we are observing from opposite ends of a base-line whose length is the diameter of the Earth's orbit; and the distance of the star will be inversely proportional to how much it *appears* to have moved between one observation and the next.

Easy to say, but the technical problems were immense. Instruments expand and warp with changes of temperature and humidity, so a change in the measured position of the star between winter and summer may be no more than an illusion generated by movements in the structure of the instrument. And there were other problems. For example, refraction — the bending of starlight as it enters the atmosphere of the Earth — affected the measures, but by how much was imperfectly understood. Newton's estimate of the distance of Sirius suggested that the apparent movement of the star every six months would be equal to no more than the width of a coin at a distance of several miles. Given the complications in making the measurement, most astronomers thought the task well-nigh hopeless.

There was however a solution, proposed as long ago as 1632 by Galileo.[39] He pointed out that if two stars lay in almost the same direction from Earth — thus forming a double star — the two stars would be *equally* affected by refraction and by any warping of the instrument, and these complications would therefore be circumvented if the apparent movement of the nearer star were measured *relative* to the more distant. For this to work, the more distant star would have to be so very far away as to be in effect a quasi-fixed reference point in the sky provided by a helpful Nature.

William hoped that double stars in his collection might prove suitable for treatment by Galileo's method,[40] and thus solve one of the great problems of stellar astronomy. But there was a snag. In spring 1782, when he submitted his first list of double stars to the Royal Society for possible publication in *Philosophical Transactions*, Maskelyne had drawn his attention to a remarkable paper by Professor John Michell of Cambridge that had appeared in 1767 (and which contains among much else the first discussion of black holes).[41] So that William could have a copy of this 31-page paper for himself, Caroline had to sit down and write it out from start to finish.[42]

Michell cast doubt on Galileo's suggested use of double stars, with an argument that pioneered the application of probability in astronomy. The Italian's method depended upon one star of the double being near and the other very distant, the appearance of a double star arising simply because the line joining the two stars happened by accident — by pure chance — to pass close to the observer in the solar system. But Michell pointed out that double stars occurred far too frequently for this to be the usual explanation. He correctly concluded that in nearly all doubles, the two component stars must be companions in space, thus producing the appearance of a double star from whichever direction they were observed. But companion stars would be lying at almost the same distance from the observer, and so useless for the application of Galileo's method.

On the other hand, if two stars were indeed companions in space, they must be bound to each other by an attractive force (or forces). Everyone would expect this force to be gravity, because Newton had claimed gravity to be universal. But his evidence was strictly limited to the solar system, and so his claim was as yet unsubstantiated. To prove that gravity did indeed operate among the stars would therefore be of the greatest interest. The proof might come from a re-examination of William's doubles, and the insights any changes might shed on the movements of the component stars in space relative to each other.

There were therefore two important reasons for monitoring changes in the appearance of a double star. If the component stars were independent, one near and the other very far away, the changes might lead to the determination of the distance of the nearer star. On the other hand, if the components were companions, measuring the changes might be a step towards confirming that the force binding the two stars together was indeed gravity. William himself was preoccupied with sweeping for nebulae, but Caroline could take over the small 20-ft and try her luck with remeasuring the doubles. She was being invited to take sole responsibility for observations of far-reaching importance.

Two decades later William himself would show that Michell was right, and that in some (at least) of his double stars the components are indeed companions in space;[43] another generation, and there would be evidence enough to show that the force binding them together was gravity.[44] A few years more, and a generalization of Galileo's method would be successfully applied to derive the first valid measures of stellar distances.[45] But William's 1783 expectations of his little sister were wholly unrealistic. Precariously perched on the ladder in the dark, Caroline would be risking life and limb anything up to 20 feet from the ground. How could she possibly hope to locate the double star, measure the relative positions of the components, and record her findings?

> Some trouble also was often thrown away during those nights in the attempt to teach me to re-measure double stars with the same micrometers with which former measures had been taken, and the small twenty-foot was given me for that purpose.... I had also to ascertain their places by a transit instrument lent for that purpose by Mr. Dalrymple, but after many fruitless attempts it was seen that the instrument was perhaps as much in fault as my observations.[46]

Fortunately for Caroline, salvation was at hand: William's early sweeps were running into trouble because of the ill-judged procedure he had adopted. Once his eyes were adjusted to the dark, he would move the great tube some twelve or fourteen degrees from side to side, and then write down what he could remember having seen. He would then have to wait patiently until his eyes were dark-adjusted once more, and repeat the procedure with the tube tilted at a slightly different elevation. It was all very unsatisfactory.

It was Caroline whose eagerness to help pointed the way out of William's difficulty. When sweeping for comets, "I generally chose my situation by the side of my Brother's instrument [the large 20-ft] that I might be ready to run to the clock or write down memorandums".[47] The solution, William realized, lay in teamwork (William would do the observing, and Caroline would do the writing) and in using the rotation of the heavens overhead to carry the nebulae into his field of view. He would direct the telescope to the south, set the tube at a given elevation, and let the sky drift past, while he watched and waited, ready to pounce on any nebula that came along. When he spotted one, he would shout out details of the observation to Caroline who was seated at a desk at an open window nearby. "[Her] care it was to write down, and at the same time loudly repeat after me, every thing I required to be written down. In this manner all the descriptions of nebulae and other observations were recorded."[48] Speed and accuracy, not understanding, were what he asked of her. As Caroline recalled many years later, "... an observer at [the] twenty-foot when sweeping wants nothing but a being that *can* and *will*

The first page of Caroline's final (1786) list of stars for use in sweeping, taken from Flamsteed's British Catalogue and arranged so as to allow her to anticipate the stars that William might next encounter. Flamsteed's folio volume was cumbersome; furthermore, he arranged the catalogue by constellation. As William swept, boundaries between constellations were irrelevant. On the other hand, all the stars he encountered were of similar North Polar Distance, and so Caroline took a writing book of convenient size and reorganized Flamsteed's stars, grouping them (without regard to constellation) first in 'zones' of similar N.P.D., and then in the order in which they passed overhead. The first coordinate (in the first two columns) she converted from Flamsteed's angles into the equivalent in time, which she found more convenient. Otherwise the numbers are as listed by Flamsteed for the year 1689, the angles in the fifth and sixth columns being related to the changes over time arising in the coordinates as a result of the wobble of the Earth's axis (specifically, they are the changes in the coordinates while the longitude of the star increases by one degree). The final column shows the 'magnitude' of the star, the brightest stars being of first magnitude. RAS MS C.2/1.2.

execute his commands with the quickness of lightning", for at times as many as a dozen nebulae would drift into the net in a single minute.[49] As a result, "In the beginning of December I became entirely attached to the writing-desk, and had seldom an opportunity after that time of using my newly acquired instrument [the 'small' 20-ft]".[50]

Clusters of nebulae were however the exception rather than the rule. In practice it turned out that the heavens rotated so slowly that there was time for the workman to raise and lower the tube slightly in an oscillating movement extending over two degrees or so, to increase the width of the strip of sky currently being swept for nebulae. Even so, the project — one of the greatest in the history of astronomy — was to take all of twenty years. From time to time William would encounter a notable star, and this would be registered and identified later from a catalogue. The positions of such stars would be used to give approximate positions for any nebulae that entered the field of view.

Each morning Caroline would make a fair copy of the night's records, and in due course she would do the necessary calculations and assemble a catalogue for publication. When she and William started to sweep, astronomers knew only the hundred or so nebulae of Messsier; when they finished, the number had increased to 2,500. It must have been chiefly for her role in these sweeps for nebulae that the distinguished French astronomer, Pierre-François-André Méchain, said of Caroline in October 1789, "Her fame will be held in honour throughout all ages".[51]

In addition to her work as William's amanuensis, Caroline was called upon to do anything else within her power to advance the great enterprise. Moving around in the dark was not without its risks. On 31 December 1783, when the new procedures had been in place for only a few days,

> about ten o'clock a few stars became visible, and in the greatest hurry all was got ready for observing. My Brother at the front of the telescope direct[ed] me to make some alteration in the lateral motion, which was done by machinery.... At each end of the machine or trough was an iron hook such as butchers use for hanging their joints upon, and having to run in the dark on ground covered a foot deep with melting snow, I fell on one of these hooks which entered my right leg about 6 inches above the knee; my brother's call "Make haste!" I could only answer by a pittiful cry "I am hooked". He and the workman were instantly with me, but they could not lift me without leaving nearly 2 oz. of my flesh behind. The workman's wife was called but was afraid to do anything, and I was obliged to be my own surgeon.[52]

The doctor later told her that a soldier with such a wound would have been entitled to six weeks' nursing in hospital. But her injury was of minor concern to Caroline: her chief worry was lest the accident might have cost William some nebulae.

> I had, however, the comfort to know that my Brother was no loser through this accident, for the remainder of the night was cloudy, and several nights afterwards afforded only a few short intervals favourable for sweeping.[53]

William and Caroline were to spend three winters in the cold and damp of Datchet. A visitor describes[54] how the sweeping proceeded:

> He has his twenty foot Newtonian telescope in the open air and mounted in his garden very simply and conveniently. It is moved by an assistant who stands below it.... Near the instrument is a clock regulated to sidereal time.... In the room near it sits Herschel's sister and she has Flamsteed's Atlas open before her. As he gives her the word, she writes down the declination and right ascension and the other circumstances of the observation.... I went to bed about one o'clock, and up to that time he had found that night four or five new nebulae. The thermometer in the garden stood at 13° Fahrenheit...,

and we may suppose it was almost as cold for Caroline sitting motionless at her desk. At times her inkwell froze. For sheer dogged perseverance, the history of astronomy knows few parallels.

Flamsteed's volumes were large format and cluttered her desk, and in them the stars were arranged by constellation rather than position. What she needed was readily-available information about the stars they might expect to encounter during the next sweep. The sweeps were of horizontal strips of sky with no more than 2° or so separating the upper and lower bounds, and so during a sweep all objects would be of roughly the same angular distance from the North Celestial Pole. She therefore took a writing-book of convenient size and compiled herself a list of the brighter Flamsteed stars. In the final version that she completed in 1786, the first group or 'zone' comprised those less than 5° from the Pole, the next those between 5° and 10° from the Pole, and so on. Within each such zone the stars were ordered in the sequence in which they passed overhead. Armed with such a list, she knew in advance the stars that William was likely to encounter next.[55]

When her help was not required by William, Caroline might sweep for comets. She thought she had found one on the night of 12 May 1784, although she was careful to describe it as a nebula.[56] To her frustration, the following night was cloudy; and on the 14th there was disappointment for her when William turned

Clay Hall, Old Windsor, from a photograph in private possession.

his 7-ft towards the object and confirmed that it was indeed a nebula.[57] She also lent a hand with some of the mirrors for the numerous instruments being made for sale or gift:

> In my leisure hours I ground 7 feet and plain mirrors from ruff ... and was indulged with polishing and the last finishing of a very beautiful mirror for S[r] W[m] Watson.[58]

The winters then were far more severe than they are now, and Datchet lies on the River Thames with its damp air. Despite the repairs to their home "there was not one room but where the rain came through the ceiling".[59] Even William's constitution began to suffer, and in June 1785 he and Caroline moved to better premises, at Clay Hall[60] on the outskirts of the village of Old Windsor. It was ideally located, being near to the Castle and on the edge of Great Park. But as William began to make improvements there, the owner, Mrs Keppel, the daughter of Sir Edward Walpole and "a litigious woman",[61] saw these as providing her with a regular excuse to increase the rent; and so, after only a year, the Herschels moved to "The Grove" at Slough,[62] now a sprawling conurbation but then a little village a couple of miles north of the Castle. The move took place on 3 April 1776. Caroline,

551 Sweep.   Breadth 2° 20'

April 3, 1786.

The telescope removed to Slough and neither my time, nor meridian ascertained.

12	9	111 17	Top. Faint moon light.		
—	16 3	112 11 / +1	9 (β) Corvi — — — — — — — —	12ʰ 23' 54",8 Cor. +7 16 3 52	112° 17' 32" 13 +5
—	30	— —	Clear and dark.		
—	50 28 / −7	111 38	45 ψ. Hydra (or 1 Hydra Continuat.) — —	12ʰ 58' 20",4 Cor. +7 50 21 59	112° 3' 1" 111 58 +5
13	0 9	112 3	46 (x). Hydra (or 2 Hydra Continuat.) — —	13ʰ 8' 6",9 Cor. +7 0 9 58	112° 7' 7" 3 +4
—	8	— —	A strong haziness in this altitude.		
—	12	111 16	Top. Left off. Not to be registered.		

552 Sweep   Breadth 2° 13'

April 15, 1786

10	48	81 39	Top. Began, a pretty strong whitish haziness.		
—	52 9	81 38	59 (c) Leonis — — — — — — — —	10ʰ 50' 20",2 Cor. −1 52 9 49	82° 49' 51" 81 38 +1 12
11	8. 34 / −8	83 8	vℑ: iℑ. for P.M. 77 (σ) Leonis p 3' 40" f 1° 29' RA 11ʰ 7' 6" 🌑 84° 22'. (1802)		
—	12. 6 { 81 39 / 82 48		77' (σ) Leonis — — — — By adjustment of the 🌑 hand.	11ʰ 10' 45",8 Cor. −1 12 6 20	82° 53' 2" 48 +5
—	20	— —	The moon appears.		
—	32	— —	The moon pretty bright.		
—	40	82 48	Top. Left off. Not to be registered.		

April 16, 1786   By Equal Altitudes Clock 20,"5 too slow.. I moved the Telescope (by Quadrant) ½ degree more east.

553 Sweep   Breadth 2° 16'

April 17, 1786.

10	14	82 46	Top. Strong twilight.		
—	32 36	84 8	35 Sextantis.		
—	40 59 / −32	83 2	cℬ. vℒ. er. 56 Leonis p 4' 19" f 0° 2' RA 10ʰ 40' 47" 🌑 83° 6. (91)		
—	41 48 / −32	85 5.	vℑ. cℑ. 56 Leonis p 4' 0" f 2° 24' RA 10ʰ 41' 36" 🌑 85° 9'. 1300		

Caroline's fair copy of the record of the first sweeps for nebulae carried out at Slough. On 17 April two nebulae are described as "Ill taken" as a result of the "blunder" of the man recruited to raise and lower the tube when instructed by William, and there is later "A stop occasioned by the same blundering

	D. M.	
45'	— —	Clear and dark.
45 48 / −32	82 40 / +1	6m. 56 Leonis — — — — — — — — — 10ʰ 45′ 35″,9 82° 45′ 16″ / 45 16 41 / Cor. + 0 20 +4
50 2 / −5	82 44	59 (c) Leonis.
51	— —	Hazy.
5	— —	Very clear.
7 4 / −18	84 17	T. b.M. 77 (σ) Leonis p 3′ 43″ f 1° 29′ RA 11ʰ 7′ 3″ PD 84° 22′ (1302)
11 −1 / −32	82 48	77 (σ) Leonis — — — — — — — — 11ʰ 10′ 45″,8 82° 53′ 2″ / 10 29 48 / Cor. + 0 17 +5
12	— —	Very hazy.
32	— —	Very clear.
49 57 / −32	85 10	7 (δ) Virginis.
0 5 / −32	83 1	11 (δ) Virginis — — — — — — — — 11ʰ 59′ 51″,0 83° 5′ 9″ / 59 33 1 / Cor. + 0 18 +.4
8 46	82 45	A Nebula. 11 (δ) Virginis f 9′ 13″ n 0° 16′ RA 12ʰ 9′ 4″ PD 82° 49′ (300)
9 26 / −32	82 59	Two. Ill taken by the blunder of the person at the handle, who is not acquainted with the business. 11 (δ) Virginis f 9′ 21″ n 0° 2′ RA 12ʰ 9′ 12″ PD 83° 3′ (301.302)
9 47	82 27	Four, The time and number is that of the last. They are scattered about. 11 (δ) Virginis f 10′ 14″ n 0° 34′ RA 12ʰ 10′ 5″ PD 82° 31′ (1425.1426.1427.1428)
— — — —		— A stop occasioned by the same blundering person.
11′ 7	83 27	A Nebula; very badly taken. 11 (δ) Virginis f 11′ 34″ f 0° 26′ RA 12ʰ 11′ 25″ PD 83° 31′ (1429)
11 34	84 22	v B. 11 (δ) Virginis f 12′ 1″ f 1° 21′ RA 12ʰ 11′ 52″ PD 84° 26′ (1430)
14 46	83 55	cB. S. 11 (δ) Virginis f. 15 13 f 0° 54′ RA 12ʰ 15′ 4″ PD 83° 59′ (1325)
16 34 / −32	84 53	p B. p L. 11 (δ) Virginis f 16′ 29″ f 1° 52′ RA 12ʰ 16′ 20″ PD 84° 57′ (96)
19 53	84 36	p T. p L. 314 Virginis of 13ᵗʰ Cat. p 16′ 39″ f 1° 43′ RA 12ʰ 20′ 13″ PD 84° 40′ (1326)
21 9	84 54	T. c L. 314 Virginis of Bode's Cat. p 15′ 23″ f 2° 1′ RA 12ʰ 21′ 29″ PD 84° 58′ (98)
22 17	83 28	cB. p L. 314 Virginis of Bode's Cat. p 9′ 15″ f 0° 35′ RA 12ʰ 22′ 37″ PD 83° 32′ (1327)
22 58	84 32	cB. c L. mbM. 314 Virginis of Bode's Cat. p 8′ 34″ f 1° 39′ RA 12ʰ 28′ 18″ PD 84° 36′ (1328)
28 51 / −15	82 51	v T. p S. 314 Virginis of Bode's Cat. p 7′ 56″ n 0° 2′ RA 12ʰ 28′ 56″ PD 82° 55′ (1329)
32 3	84 54	T. p S. 314 Virginis of Bode's Cat. p 4′ 29″ f 2° 1′ RA 12ʰ 32′ 23″ PD 84° 58′ (1330)
37 5 / −33	82 53	6m. U.¹⁰⁰ N: 314 Virginis of B's Cat. 12ʰ 36′ 52″,2 82° 57′ 1″ / 36 32 53 / Cor. + 0 20 −+4
		L.

person". Thereafter things improve. The numbers in parentheses are the serial numbers subsequently assigned to the nebulae by Caroline in the 1820s. RAS MS W.2/3.5.

in an oft-quoted passage, makes the romantic claim that "the last night at Clay Hall was spent in Sweeping till daylight, and the next the Telescope stud ready for observation at Slough". In fact the last Clay Hall sweeps took place on 28 March.[63] And ready the telescope might have been on the evening of 3 April, but William had not yet determined the precise direction of south, and the hour's sweep was "not to be registered". Nor was the next sweep, which did not take place until the 15th. On the 17th the problem was not in the technology but in the incompetence of the man newly hired to raise and lower the tube during sweeps: "Two [nebulae]. Ill taken by the blunder of the person at the handle", though fortunately Caroline later found that these nebulae had already been observed on a previous occasion. An hour later, "A stop occasioned by the same blundering person". But thereafter matters improved.[64]

Slough High Street ran east–west, and along it passed the coaches between London and Bath. It was crossed by a road leading to Windsor from the north, and the Crown Inn stood at the southeast corner of the cross-roads. Two hundred yards away to the south, on the east side of Windsor Road, was "The Grove",[65] and this was to be William's home for the rest of his life. It had two large and two small bedrooms and a loft for servants.[66] The stables had been converted into a dwelling, and it was there that Caroline had her quarters. The harness room was turned into a writing room, and a small staircase led out onto the flat roof where she could set up her comet sweeper.[67]

Even in the early months at Datchet, Caroline had become increasingly concerned about their finances. The expenses involved in the move from Bath had been considerable, and William's modest pension was paid in arrears. He had had to draw on his savings to fund the construction of the large 20-ft on which his observing programme depended. The instrument needed further improvements, but these

> could not be supplied out of a salary of 200 pound per year; especially as my Brother's finances had been too much reduced, — during the six months previous, before he received his *first* quarterly payment of *fifty pounds* which was Michaelmas 1782 — by traveling from Bath to London, Greenwich, Windsor backwards and forwards, transporting the Telescope [to and from the Castle] &c. Breaking up his establishment at Bath and forming a new one near the Court....[68]

The King, she felt, had been cavalier in assuming that if William's pension was inadequate, he could make up the shortfall by manufacturing telescopes for sale. She suspected, perhaps rightly, that George imagined William had little else to

Windsor Road, Slough, from a photograph in private possession. Extreme left, the Crown Inn, and next to it the cottage occupied by Mary's mother. William's residence was further to the south (right), some 200 yards from the Inn.

do in the daytime; and surely there would be no lack of orders, for all the world knew that William's telescopes were without equal. Any gentleman — indeed, any monarch — would be proud to have one that he could display to his guests.

Caroline would have us believe that William resented this drain on his time and energy. He was already in middle age, "and felt the injustice he would be doing to himself and to the cause of Astronomy by giving up his time to making telescopes for other observers".[69] Had William realized just how little use was to be made of his instruments, he might well have had misgivings.[70] But Caroline mistakenly imputes her own sentiments to her brother, overlooking the satisfaction it gave him to be pre-eminent in a profession so far removed from the "fiddling" he had been taught as a boy. There is no mistaking the exhilaration in his Datchet letter of 10 March 1785 to Alexander, when his production of telescopes was at last in full swing:

> I have bought a compleat set of tools for working in Brass; erected a small forge; have a Brass workman; the Cabinet maker is imployed; the Joiner at

work; the Smith forging away, so that I hope to get some instruments finished pretty fast.[71]

He supplied price-lists to intending customers, and invited orders for instruments of all sizes, practicable or otherwise. A 7-ft cost 100 or 150 guineas, depending on the size of the mirror. At the other extreme, he offered a 40-ft reflector at no less than 8,000 guineas, but (perhaps fortunately) it drew no customers.[72] William never lost the sense of achievement when he packed up a new telescope and despatched it to its new owner (complete with instructions written in Caroline's clear script): "... by this traffic Dr. Herschel established his fame as one of the greatest mechanics of his day, and also set himself at ease in pecuniary matters."[73] Indeed he would go on making telescopes long after there was any financial pressure on him to do so.

Alexander continued to be his enthusiastic collaborator in telescope-building, for the journey from Bath was an easy one and in the summer Alexander had time on his hands. Indeed, William tried at first to set up Alexander in business on his own. In 1785 he told the German astronomer Johann Bode: "You enquired, by way of Monsieur Zach, if I could furnish Newtonian telescopes for amateurs and what would be the price. My brother now makes them of every size." The following year, Watson enquired after William's "design of setting up your brother to make [reflectors] for sale", and in 1787 the Italian astronomer Barnaba Oriani wrote in his diary that "his brother makes telescopes for sale when not engaged as a musician in Bath". But there is in fact no record of any such sale, and it seems likely that Alexander was one of those who find satisfaction enough in the solution to technical challenges.[74]

But funding the large 20-ft was only the start of the problems facing William in his ambitions to study 'the construction of the heavens'; the exploration of the furthest depths of space called for nothing less than the largest telescope that his unrivalled ingenuity could contrive. William thought it would be possible to scale up the 20-ft with its 18-inch mirrors until it became a 40-ft with 48-inch mirrors. This was taking technology into uncharted realms, and the expenses involved could only be guessed. Certainly they would equal a king's ransom, and for William the only way forward was to hold King George to ransom.

William's old ally from Bath, William Watson, had visited him at Datchet. Watson had been outraged from the start at the small pension the King had awarded William, and now he found the once-prosperous musician had become a (relatively) impoverished astronomer. It was too late to reopen that question; but perhaps

Watson could help bring about the great new instrument on which William had set his heart. Watson

> saw my Brother's difficulties and expressed great dissatisfaction. And on his return to Bath he met among the Visitors there several belonging to the Court ... to whom he gave his opinion concerning his Friend and his situation very freely. In consequence of which, very soon after, my Brother had, through Sir J. Banks, the promise that 2000 pounds would be granted for enabling him to make himself an Instrument.[75]

Banks had long been one of William's greatest admirers, and as President of the Royal Society he had immense influence. The exact sequence of events is uncertain;[76] but it seems probable that William had a preliminary audience with the King, and in his enthusiasm manipulated matters so well that the King was left with the impression that it was he who had opted for an ambitious 40-ft, rather than a more realistic — and cheaper — 30-ft: "... having proposed to the King either a 30 or a 40 feet telescope, His Majesty fixed upon the largest."[77]

It was to prove a disastrous choice. The 25-ft reflector with 24-inch mirrors that William made for the King of Spain at the turn of the century was to be his finest achievement as a telescope builder, and it was with reluctance that he parted with it to its new owner.[78] A 30-ft with (say) 36-inch mirrors might have been feasible;[79] the 40-ft with 48-inch mirrors was to prove several steps too far.

It seems the King agreed that William should write Banks a formal application for funds (which Banks would then pass to the King with his endorsement), setting out the possibilities that would be opened up by such a telescope, and the financial implications. This letter is now in the royal archives. William explains that the construction of the larger 20-ft "has pretty nearly exhausted the little stock I had collected from the profits of my former musical avocation at Bath. It is true I might collect some small sums by making and selling a few telescopes", but life is short and it would be self-defeating for him to waste the years remaining to him "in laborious, mechanical operations".

> ... the telescope I would wish to undertake, should be of the Newtonian form, with an octagon tube 40 feet long and five feet diameter; the specula [mirrors], of which it would be necessary to have at least two, or perhaps three, should be from 36 to 48 or 50 inches in diameter.[80]

The enclosed Estimate of Expenses[81] totals £1395, with annual running costs of £150. Much the largest item is £500 for a pair of mirrors each to weigh rather over

A little-known (and unfinished) painting of the Herschel home in Slough, by John Gendall (1789–1865). The 40-ft is shown on the left. Reproduced by permission of the Buckinghamshire Archaeological Society.

half a ton. One reason why the project eventually ran well over budget was that the first mirror, weighing a little under half a ton, was so thin that it was distorted in use by its own weight. To avoid this defect in the second (and serviceable) mirror, William had to use nearly a ton of expensive metal. He was later to cost this second mirror alone at £500.[82]

In September 1785 the King duly made what he and Banks no doubt considered a realistic and once-for-all grant: £2000, equalling the capital sum requested plus the running costs for the first four years. The project was a mammoth undertaking. The mirrors, along with other major apparatus, had to be shipped up river from London. Whole teams of artisans of all kinds were recruited to work on site;[83] irritatingly, these passed the time by "talking and sometimes singing on all sorts of subjects",[84] making it difficult for William to concentrate. Yet the work could not be allowed to interfere with the sweeps for nebulae:

> If it had not been sometimes for the intervention of a cloudy or moon-light night, I know not when my Brother (or I either) should have got any sleep; for with the morning came also his work-people of which there were not less than between 30 or 40 at work for upwards of 3 months together, some employed with felling and roothing out trees, some digging and preparing the ground for the Bricklayers who were laying the foundation for the Telescope, and the Carpenter in Slough with all his men. The Smith meanwhile was converting a wash house into a Forge, and manufacturing complete sets of Tools required for the work he was to enter upon; and many expensive tools were furnished by the Ironmongers in Windsor as well for the Forge as for the turner of brass man.[85]

King George had no conception of the scale of the task William had set himself, and this would spell trouble in the years ahead as work proceeded — unduly slowly in the eyes of the impatient sponsor — and as expenses mounted. As early as November 1785, only weeks after making the grant for the construction, the King was looking for tangible signs of progress. Watson wrote to William:

> S[r] Joseph Banks is come to Town, & expressed a wish to know from you what preparations you have made relating to the great Telescope, & how far you have proceeded in the work itself. He said that he was very desirous of knowing, that he might be enabled to give the King a history of your proceedings.[86]

The King's total lack of grasp of what was entailed in the construction was illustrated the following year, when George asked William to drop everything and go

One of the five 10-ft reflectors commissioned by the King at the outset of William's career as a professional telescope-maker. It was presented to the fourth Duke of Marlborough and was housed in one of the towers of Blenheim Palace; it is now in the Whipple Museum of the History of Science, Cambridge University. Others went to the King's observatory at Kew, Göttingen University, and Count de Brühl. In 1791 the King had yet to give instructions as to the destination of the fifth instrument, which suggests that his motive had been primarily to launch William on his career.

in person to Göttingen, and present the University there with one of the 10-ft reflectors William was making for him. William must have been near despair at the interruption to the construction of the great telescope (to say nothing of the loss of precious nights in which to sweep for nebulae). But he could hardly refuse his royal benefactor, and it would give him a chance to visit his family in Hanover.

Alexander came up from Bath to accompany him. On their move to Datchet, William and Caroline, fearful of what their brother would make of his life when left to his own devices, had tried to persuade him to leave Bath and settle in London, to be near them. Jacob for similar reasons had invited Alexander to accept a post in the orchestra in Hanover. All in vain: "before we saw him the next year he was married (wretchedly) and we saw him never otherwise but discontented after this our separation."[87] Alexander brought his wife to stay with Caroline while he was abroad, and Caroline no doubt did her limited best to conceal her disapproval of her sister-in-law; to her diary she confided, "I was obliged frequently to sacrifice an hour to her gossipings".[88]

William and Alexander were abroad for most of July and August 1786. Caroline did her best to keep things ticking over. It was not always easy, "for I could only look upon myself as an individual who was neither Mistress of her Brother's house nor of her time".[89] On 8 July, Caroline

> paid the Smith and the Gardener who worked 5 days this week. N.B. I had a deal of trouble with this fellow; for my orders were not to employ him more than 3 days just to keep the grass-plot in order...; but he would be idling about the premises and gave me the name of stingy in the village, because I objected to his being there when he was not wanted.[90]

29 July:

> I paid the Smith. He received to-day the plates for the 40 ft. Tube; they are above half of them bad, but he thinks there will be as many good among them as will be wanted.[91]

She was constantly plagued with time-wasting visitors, come to see the 20-ft and the progress of the 40-ft. Her productive hours were spent in preparing for publication a catalogue of nebulae and clusters of stars, for that spring the total discovered by the Herschel team had reached one thousand, ten times the number previously known to astronomers.

Three days later she had an exciting discovery to record. 1 August: "I have calculated 100 nebulae today, and this evening I saw an object which I believe will prove to-morrow night to be a comet." 2 August: "To-day I calculated 150 nebulae. I fear it will not be clear to-night, it has been raining throughout the whole day, but seems now to clear up a little. 1 o'clock; the object of last night *is a Comet*. I did not go to rest till I had wrote to Dr. Blagden [secretary of the Royal Society] and Mr. Aubert to announce the comet."[92]

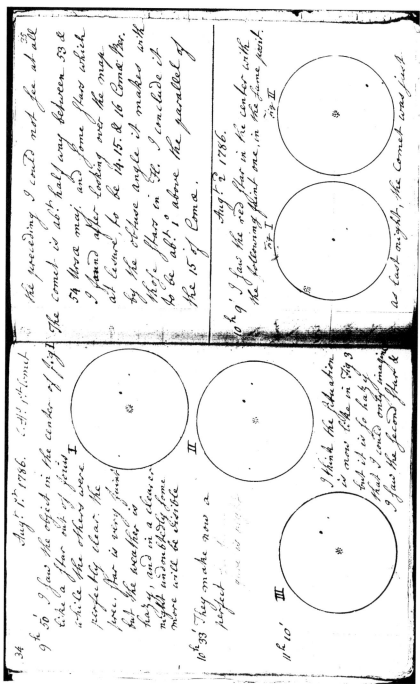

Pages of Caroline's Book of Observations recording the discovery of her first comet. RAS MS C.1/1.1, 34–35.

Aubert replied on the 7th:

> I wish you joy most sincerely for the discovery. I am more pleased than you can well conceive that *you* have made it and I think that your *wonderfully clever* and *wonderfully* amiable Brother, upon the news of it, shed a tear of joy. You have immortalized your name and you deserved such a reward from the Being who has ordered all these things to move as we find them, for your assiduity in the business of astronomy and for your love for so celebrated and so deserving a brother.[93]

Blagden, who had wide contacts in the English scientific community, told her:

> I believe the comet has not yet been seen by anyone in England but yourself. Yesterday the Visitation of the Royal Observatory at Greenwich was held, where most of the principal astronomers in and near London attended, which afforded an opportunity of spreading the news of your discovery, and I doubt not but many of them will verify it the next clear night. I also mentioned it in a letter to Paris, and in another I had occasion to write to Munich in Germany. If the weather should be favourable Sunday evening, it is not impossible that Sir Joseph Banks, and some friends from his house, may wait upon you to beg the favour of viewing this phenomenon through your telescope.[94]

And so it happened. On 6 August the President and the Secretary of the Royal Society, together with Lord Palmerston, journeyed to Slough expressly in order to look through Caroline's telescope and see Caroline's comet.[95] William on his return from Germany was summoned by royal command to demonstrate the comet to the Royal Family. The novelist Fanny Burney was present. "The comet was very small, and had nothing grand or striking in its appearance; but it is the first lady's comet, and I was very desirous to see it."[96] To Caroline the scullery of her childhood home in Hanover must have seemed a long way away.

A year later, on 17 August 1787, she and William were to welcome to Slough a party that included the King, the Queen, the Duke of York, the Princess Royal, Princess Augusta, the Duke of Queensbury "and a company of Lords and Ladies", ostensibly to examine the tube of the great telescope while it was still on the ground. The occasion was a great success:

> And the Archbishop of Canterbury following the King and finding it difficult to proceed, the King turned to give him the hand, saying: "Come, my Lord Bishop, I will show you the way to Heaven."[97]

It was surely no accident that the royal visit took place when it did, for the great project was in financial crisis. On 18 July a worried William wrote to a member of the royal household:

> In a letter which Sir J. Banks laid before his Majesty [in 1785], I have mentioned that it would require 12 or 15 hundred pounds to construct a 40-ft telescope, and that moreover the annual expences attending the same instrument would amount to 150 or 200 pounds. As it was impossible to say exactly what sum might be sufficient to finish so grand a work, I now find that many of the parts take up much more time and labour of workmen, and more materials than I apprehended they would have taken, and that consequently my first estimate of the total expence will fall short of the real amount.

Would the King prefer William to make another application through Banks, or would he prefer William "to continue the workmen and apply from time to time for such sums as may be wanted"?[98]

Not for the last time in the history of astronomy, an astronomer seeking support had been modest in his initial demands, knowing that the funding body, confronted later with a choice between writing off all the money spent so far or coughing up more, would cough up. The message to the King was clear: pay up, or else....

Indeed, William's letter assumed — with barely a hint of apology — that the King would indeed pay up, and merely enquired whether he preferred to do so by lump sum or by payments against receipts. The King however had been under the impression that he had already paid the full cost of the telescope. For all his affability towards the Archbishop, he was not amused.

The royal visit to Slough bears all the marks of a covert enquiry on the part of the grant-giving body, as to how well past monies have been spent and whether a supplementary grant would be justified. It would seem that the King departed leaving William under the impression that he had received the royal equivalent of the thumbs up, and that he could now pen a formal letter to Banks, the established intermediary. It begins:

> By the King's great liberality and encouragement I have been enabled to bring the 40 feet telescope to the advanced state in which you have seen it; and as it is his Majesty's intention further to support the construction and completion of this instrument as well as to provide for such necessary annual expenses as will be connected with its being kept up and serving for a series of observations, I shall lay before you an account of the things which are still wanting, with an estimate of the expenses they may occasion.

The 40-ft reflector, with the huts that housed Caroline and the workman. This is not the familiar engraving that accompanied William's account of the 40-ft in *Philosophical Transactions* in 1795, nor is it the (rare) engraving he published in February 1794, but the watercolour (in private possession) on which these engravings were presumably based.

Perish the thought, he continues, that he is using this as an excuse "to enrich myself by his Majesty's bounty", though the idea does seem to have crossed the royal mind. On the contrary, William has "no other view than the advancement of astronomy, the honour of a liberal Monarch, and the glory of a nation which stands foremost in the cultivation of arts and sciences". The cost of completing the telescope will be £950, assuming that no disaster befalls the great mirror (which he prices at £500).[99]

He now comes to the delicate matter of running costs:

But the annual expence, which you find Sir, by the particulars I have given, amounts to 245 pounds, and which by one of his Majesty's pages I have mentioned to the King, I would endeavour by good management to reduce to 200 pounds, I fear will leave one article unprovided for which it will be proper to mention.

You know Sir, that observations with this great instrument cannot be made without four persons: the Astronomer, the assistant, and two workmen for the motions. Now, my good, industrious sister has hitherto supplied the place of assistant, and intends to continue to do that work. She does it indeed so much better, to my liking, than any other person I could have, that I should be very sorry ever to lose her from that office. Perhaps our gracious Queen, by way of encouraging a female astronomer, might be enduced to allow her a small annual bounty, such as 50 or 60 pounds, which would make her easy for life, so that, if anything should happen to me she would not have the anxiety upon her mind of being left unprovided for. She has often formed a wish but never had the resolution of causing an application to be made to her Majesty for this purpose; nor could I have been prevailed upon to mention it now, were it not for her evident use in the observations that are to be made with the 40 feet reflector, and the unavoidable encrease of the annual expences which, if my Sister were to decline that office would probably amount to nearly one hundred pounds more for an assistant.[100]

William was asking for under £1000 for additional capital costs together with running expenses of £200 (plus £50 for Caroline, presented as cheap at the price). Surprisingly, the King in fact gave him all this and another £1000 besides. One might have suspected that Banks, with his unrivalled experience of fund-raising, had discreetly taken the opportunity to increase the sum requested, so as to create a contingency fund; but the letter in the royal archives is exactly as William (or, more exactly, Caroline) wrote it.

Rare is it that a funding body gives *more* than requested. Despite being told that his original grant of £2000 was a write-off unless huge further sums were forthcoming, the King not only gives the additional sum requested but doubles it. One might suppose he was delighted with the progress of the enterprise.

In fact he was furious. A seemingly-chastened Banks sent a terse reply to William on 23 August. The complete letter reads:

I have this moment seen the King who has granted all you ask but upon certain conditions which I must explain to you. Will you be so good as to come to me in Soho Square tomorrow as soon as convenient that we may finished this matter & that I may report to him before he sets out for Windsor.[101]

William had the ability to wipe unhappy memories from his mind, destroying the evidence as he did so; Caroline never forgot, but she too destroyed the evidence. And so it is that of this, one of the most unpleasant episodes in their lives, almost nothing survives save a consoling letter from Watson to William, and a couple of sentences Caroline penned for family eyes only, forty years after the event:

> But there can be no harm in telling my own dear nephew, that I never felt satisfied with the support your father received towards his undertakings, and far less with the ungracious manner in which it was granted. For the last sum came with a message that more must never be asked for. (Oh! how degraded I felt even for myself whenever I thought of it!)[102]

Watson's long letter, dated 7 September, could be summed up in the two words, "Never mind!":

> I do most sincerely sympathize with you, & feel in some measure as you must feel at the unworthy treatment you (& I may add Science) has received. But I sincerely hope by the latter part of your letter that the Storm is past, & that the K— is brought by reflexion to know you ... a little better. For I am confident that no prince could have secured so much reputation at so small an expence, as has done his Majesty by the countenance he has shewn you.
>
> Let me hope, my dear Sir, that this affair has ceased to give you inquietude, & has not lessened your zeal for Science. Remember you have much cause of comfort & even of exultation. By your great discoveries, mechanical improvements & learned communications, far superior to any that has ever fallen to the lot of one person to make, in your line before you; you have gained a high & universal reputation...

and if William is pushed for cash, Watson will be happy to send him a couple of hundred pounds.[103]

In later life Caroline was to blame the falling out on the King's "*shabby, mean-spirited advisors* who were undoubtedly consulted on such occasions" (though she explicitly exempts Banks from this charge).[104] Yet the royal wrath must have been impressive indeed if the grant to William of everything he had asked for, and an extra thousand pounds besides, constituted "unworthy treatment". The King was in fact only a year away from his first attack of "madness" (probably porphyria) and this may have contributed to the violence of his outburst.[105]

Relations between the King and his astronomer never recovered. William received no public honour from George III — his richly-deserved knighthood was not conferred until 1816,[106] by which time the severity of the King's illness had resulted in his son's taking control as Prince Regent; and it was the Prince Regent, now George IV, who in 1820 doubled William's modest pension[107] (and confirmed Caroline's).

From now on William's expenditure was closely monitored. Early in 1790 he writes a formal letter (probably to Banks): "Dr. Herschel will prepare without delay an account of the moneys already receiv'd from the King towards the construction of his 40 feet telescope and lay it before Sir Jos: Banks. He has every reason which opinion founded on experience can give, to be certain that the additional money he has ask'd will be sufficient to finish the instrument; nothing but unforeseen accidents can make it otherwise."[108] But the immediate purpose of the letter is to advise the King that the telescope is up and running, and therefore he "humbly hopes his Majesty will consent to his annual allowance for expences commencing immediately". Accompanying the letter is a list of anticipated annual expenses in extraordinary detail. Twelve men engaged in polishing for six weeks are expected to consume one pint of beer per day and this will total £3. 12s. "One man in livery to attend company and strangers that wish to see the telescope in the day time and help to work the instrument during the night", £35. Winter time, "Two fires and four or five candles all night, when the weather is fine. One fire day and night the whole winter. In the day for company that comes to see the telescope, in the night waiting the coming of stars tho' the weather should be cloudy", no actual cost stated but in 1787 he had quoted £15 for this.[109] William is having to account for every last penny.

"Degraded" though Caroline might imagine she had felt, as she looked back forty years after the event, at the time she was overjoyed to have money that she could call her own:

> ... in October I received £12.10, being the first quarterly payment of my Salary. And the first money I ever in all my lifetime thought myself to be at liberty to spend to my own liking. A great uneasiness was by this means removed from my mind, for though I had generally (and especially during the last busy 6 years) been the keeper, almost, of my Brother's purse, with a charge to provide my personal wants with only annexing in my accounts the memorandum: *"for Car."* to the sums so laid out, they did when cast up hardly amount to 7 or 8 pounds per year since the time we had left Bath. For nothing but Bankruptcy had all

An engraving of Upton House, home of John Pitt and his wife Mary when William and Caroline first settled in Slough.

the while been running through my silly head, when looking at the sums of my weekly accounts, [though] knowing they could be but trifling to what had been and was yet to be paid in Town [for the construction of the 40-ft].[110]

There was a further and compelling reason for the application to the King for a salary for Caroline: after nearly half a century of bachelorhood, William was planning matrimony.

For Caroline this revived an old nightmare. As a teenager back in Hanover, acting as housemaid for her brother Jacob, Caroline had worried over what would become of her when he married and no longer needed her services. With William, since Caroline would not have known of his attempted courting of Elizabeth Harper, there had hitherto been no cause for any such concern. True, Bath gossip had linked him with the rich widow Mrs Colebrook, but nothing he did suggested there was substance in this. As Fanny Burney remarked, William Herschel "seems a man without a wish that has its object in the terrestrial globe".[111]

In 1786, however, William became accustomed to visit two neighbours, John and Mary Pitt, whose comfortable home[112] at the neighbouring village of Upton was a short walk across the fields and whose company he enjoyed. Indeed, they were more than neighbours, for William and Caroline's house belonged to Mrs Pitt's mother.[113] She lived close by, in a little house situated between the Herschel home on Windsor Road and the Crown Inn on the corner of Windsor Road and the

High Street. She was the widow of Adee Baldwin, a wealthy merchant in the City of London who had owned considerable property around Slough,[114] including the Crown Inn and the land between it and the Herschel home.[115]

John Pitt was then in poor health, and he died at the end of the summer of 1786. In his will he left two thousand pounds to his son Paul Adee Pitt, which he was to inherit when he came of age.[116] This sum alone amounted to ten times William's annual pension. He left a life-interest in the rest of his estate to his thirty-six-year-old widow Mary;[117] it was then to pass to Paul, though as it happened Mary survived him.[118] John Pitt's will does not put a total value on his estate and so we do not know exactly how much his widow inherited; but his possessions included the Dolphin and other property in the nearby parish of Langley Marsh. Mary was more than comfortably provided for, with a legacy we might conservatively estimate at ten thousand pounds. All this was of course in addition to the inheritance — perhaps of a similar amount — she would one day receive from her mother.[119] In her diary Fanny Burney was to write of her: "... she was rich too! and astronomers are as able as other men to discern that gold can glitter as well as the stars."[120]

A neighbour, Mrs Papendiek, records that Mary,

> poor woman, complained much of the dullness of her life, and we did our best to cheer her, as did also Dr Herschel, who often walked over to her house with his sister of an evening, and as often induced her to join his snug dinner at Slough.
>
> Among friends it was soon discovered that an earthly star attracted the attention of Dr Herschel. An offer was made to Widow Pitt, and accepted. They were to live at Upton, and Miss Herschel at Slough, which would remain the house of business.

Under these arrangements, Caroline would have continued as housekeeper at Slough.

> All at once it struck Mrs Pitt that the Doctor would be principally at the latter place, and that Miss Herschel would be mistress of the concern, and considering the matter in all its bearings, she determined upon giving [the engagement] up. Dr Herschel expressed his disappointment, but said that his pursuit he would not relinquish; that he must have a constant assistant and that he had trained his sister to be a most efficient one. She was indefatigable, and from her affection for him would make any sacrifice to promote his happiness.[121]

Given the choice between astronomy with Caroline and marriage to the Widow

Pitt, William had chosen astronomy with Caroline. But he was prepared to negotiate. In the autumn of 1787, Mrs Papendiek wrote:

> A few days after our return home Mrs Pitt called to tell us that the offer from Dr Herschel had been renewed, and again accepted, under the following arrangement. There were to be two establishments, one at Upton and one at Slough; two maidservants in each, and one footman to go backwards and forwards, with accommodation in both places, and Miss Herschel to have apartments over the workshops.[122]

Caroline had been sacrificed.

One might have expected both bride and groom, neither of them young, to be conscious of the ticking of their biological clocks, but plans for the wedding proceeded at a leisurely pace. William was worried in case astronomers thought he would be neglecting his duties if he spent some nights with his wife instead of with the stars. Watson was instructed to take soundings.

He replied on 24 March 1788, six months after the engagement had been finalized. William's friends were, it seemed, on the whole favourable, though some had expressed misgivings:

> I had no opportunity when I was last with you, of speaking a word apart with you, or I should have given you the result of what I have collected relating to the general opinion of your friends upon your future marriage. And I am extreamly happy to inform you, that excepting some little fears with respect to Astronomy, I have not heard anything which you would have disliked; on the contrary, it seemed to meet with the approbation of everyone.

Indeed, for his part, Watson thought that if William adopted a less intensive and stressful life-style it might in fact benefit astronomy in the long run.[123] But what, we might ask, would William have done if his friends had been opposed? Would he have broken off the engagement a second time? This hardly seems likely.

It is clear that for Caroline this was the disaster she had long dreaded.

> And the 8th of that month [May 1788] being fixed on for my Brother's marriage it may easily be supposed that I must have been fully employed (besides minding the heavens) to prepare everything as well as I could, against the time I was to give up the place of a Housekeeper.[124]

She and Alexander were witnesses to the wedding, which took place at Upton Church. Sir Joseph Banks was best man, and

Miss Herschel received them at Slough, which was the honoured house for the reception of the newly married pair and where they spent their honeymoon.[125]

Caroline was now to take second place in William's affections. But this was not all. For sixteen years she had been his housekeeper, and in charge of the expense account. What she needed for her personal expenses she had taken, for all the world as though she was his wife. Now she was redundant, and without financial support. William, whose monetary worries would end with his marriage, had felt able to offer her the equivalent of a retirement pension, in the manner of all good employers. But to Caroline this would have made her dependent upon fraternal handouts, mirroring her dependence on Jacob when a young woman in Hanover long ago.

She had therefore proposed the alternative that William put to the King when submitting his estimate of running expenses for the 40-ft. Marriage would not alter his need for an astronomical assistant. Until now she had worked in return for her board and lodging; should she not receive a salary in lieu?

I refused my dear brother's proposal (at the time he resolved to enter the married state) of making me independent, and desired him to ask the King for a small salary to enable me to continue his assistant. £50 were granted to me, with which I was resolved to live without the assistance of my Brother.[126]

Even if "A great uneasiness was by this means removed from my mind",[127] Caroline did not regard the award as over-generous. £50 was "exactly the sum I saved my Brother at Bath in writing music" — and she had done this copying in comfort, "by a clean fire side".[128]

Caroline was soon to find that royal promises were one thing, laying hands on the cash another. She was later to destroy her Journals for the next nine years,[129] thereby erasing from the record her bitterness over William's marriage, and she ended her autobiography with the day of his wedding; and so we are poorly informed about this period of her life (which included the death of her mother[130] and the puzzling murder "in the field" of Jacob[131]). But a letter she wrote after William's death reveals that she soon found it impossible to get the King to honour his commitment to her.[132] Arrears eventually amounted to no less than nine quarters' worth, and Caroline was penniless.[133] William had insisted that if she was ever in financial difficulties it was to him that she must turn, so this she did. William was more experienced in twisting the royal arm. Her payments were resumed; and Caroline had financial independence at last.

4

Internal Exile

On 9 March 1792 Mary Herschel gave birth to a son, John Frederick William. Gradually, Caroline began to rebuild her relationships. The newly-arrived infant served as a focus for his aunt's love; and his mother, described by Fanny Burney's father, Dr Charles Burney, as "sensible, good-humoured, unpretending, and obliging",[1] proved an affectionate sister-in-law. To William, Mary was a loyal and loving wife, though a disappointment as a telescope maker: "Mrs Herschel polished a small speculum for a 7 feet Newtonian.... The emery at last not being very fine it took a long time to give it a good gloss."[2]

As Watson had foreseen, the married William pursued his researches as before, but at an altogether more measured pace. The Herschel family took regular and sometimes extended holidays. Mary Herschel's niece Sophia, "a sweet timid aimiable girl",[3] frequently went with them; Caroline, never. She was left behind to keep an eye on the telescopes, receive visitors, and do whatever she could find to keep herself in useful employment and so justify her salary as assistant to William. Alexander, whose wife had died in 1788,[4] would often come from Bath and keep her company.

On these holidays, William and his family lived in style. In 1809 they made an eight-week tour of the north of England that cost them £188 17s, almost the whole of William's pension for the year. When he set off in July 1810 William had no less than £400 in his cash box, and this scene was repeated a year later.[5]

Meanwhile money poured into Mary's bank account from all directions. In January 1795 her "Aunt Clark" died, leaving jointly to Mary and John no less than £5,000 of 3% stock; in addition Aunt Clark left Mary a half-share in her house at Walton, together with the Crown Inn in the same village (on which Mary later realised £330), and property in Greenwich. Three years later came the death of Mary's mother; unfortunately we do not know the total value of her very considerable estate, merely that Mary received £185 as her share of her mother's cash and furniture. Mary at some stage acquired a farm, whose stock she sold in 1809 for £761.[6] Money was no object, William's annual pension of £200 an irrelevance, and his manufacture of telescopes for sale motivated by something other than financial need.

Despite her fierce sense of independence, Caroline permitted herself a small share in this increased prosperity, though not without certain misgivings. As she later wrote to Mary, "In 1803, you and my brother insisted on my having £10 quarterly added to my income, which I certainly should not have accepted if I had not been in a panic for my friends in Hanover, which had just then been taken by the French".[7] From then on, "Miss H—, £10" occurs every quarter in Mary's accounts.

Sometimes William's holidays were pure relaxation, but at other times they took him to the industrial heartland of Britain. As befitted "one of the greatest mechanics of his day",[8] he took an intense interest in the machinery that was powering the Industrial Revolution, and his travel diaries are full of elaborate drawings that he uses to illustrate his manuscript notes on how the machinery operated. Indeed, in 1793 he was a witness for James Watt in the case of Watt *v.* Bull.[9]

He undertook a journey of a quite different nature in the summer of 1802, soon after the Peace of Amiens made it possible for visitors from England to travel to Paris for the first time in a decade. There he met on a number of occasions with the great exponent of the mechanics of the solar system, Pierre-Simon de Laplace; and they had much to discuss, for Laplace held that the planets had condensed out of a nebula through the action of gravity, while William believed the same of stars. William's eyebrows were raised at Laplace's wife: "His Lady received company abed; which to those who are not used to it appears very remarkable."[10]

The highlight of the visit was his audience with Napoleon, the "first Consul", who on entering the room from the garden took a seat and invited William to do likewise; but as everyone else remained standing, William thought it best to bow his thanks and stay on his feet.

> The first Consul then asked a few questions relating to Astronomy & the construction of the heavens to which I made such answers as seemed to give him great satisfaction. He also addressed himself to Mr Laplace on the same subject, and held a considerable argument with him in which he differed from that eminent mathematician. The difference was occasioned by an exclamation of the first Consul, who asked in a tone of exclamation or admiration (when we were speaking of the extent of the sidereal heavens): "And who is the author of all this!" Mons. De la Place wished to shew that a chain of natural causes would

"... sensible, good-humoured, unpretending, and obliging." Mary Herschel in 1805, from a miniature painted on ivory by J. Kernan, in private possession.

account for the construction and preservation of the wonderful system. This
the first Consul rather opposed. Much may be said on the subject; by joining
the arguments of both we shall be led to "Nature and nature's God".[11]

Napoleon then tactfully shifted the conversation to the breeding of horses in
England; and before long delicious ices were being offered round.

While in Paris William also took the opportunity to visit Charles Messier, Caroline's great rival as a comet-hunter, who was now in his seventies. It was over
twenty years since he had had a fall into an ice-cellar, but he was still suffering
from the consequences. William took to him: "Merit is not always rewarded as it
ought to be."[12]

When at home William continued his observations, albeit at an altogether less
hectic pace. Seventeen days after the wedding William and Caroline were once more
sweeping for nebulae. They already had almost enough materials for the second
catalogue of one thousand nebulae, having multiplied twenty-fold the number of
known nebulae in only five years. However, the completion of the sweeps, and the
compilation of the final catalogue of just five hundred nebulae, would take them
until 1802, another fourteen years. Some of the remaining sweeps were of sky
between the zenith and the North Celestial Pole, regions where it was awkward
and time-consuming to manoeuvre the telescope; and William was distracted by
the interests he had recently developed in other observational questions — he
was for example compiling catalogues of the comparative brightnesses of stars,
as aids for the detection of stars that varied in brightness. But there is no denying
that for him, the days of frenetic activity, driven by the Protestant work ethic that
had been instilled into them as children, were over.

This would not have been evident to the outside world: in terms of the sheer
volume of his scientific publications, the turn of the century marks the halfway
point of William's career, though many of his later papers were devoted to the
interpretation of data assembled years before. But two especially important achievements lay immediately ahead.

He had long been fascinated with the physical structure of the Sun, for he believed
in the Principle of Plenitude, according to which God had not restrained himself
avoidably (so to speak) when creating the universe. As a result, rational beings
were to be found everywhere, and among these were inhabitants of the Moon and
the Sun. While still resident in Bath William had unwittingly flouted convention
by mentioning lunar inhabitants in his paper for *Philosophical Transactions* on the
mountains on the Moon; and he was equally convinced that the Sun was "a most

magnificent habitable globe".[13] To shield the inhabitants from the outer shell of fire, there was an inner shell of cloud, which we glimpse as sunspots. Persuading other astronomers of this implausible scenario would not be easy, and so William devoted a lot of effort to observations of the surface of the Sun, during which he protected his eyes with coloured glass shades.

In his attempts to determine the most satisfactory form of shade, he was surprised to find that the sensation of heat seemed to depend more on the colour of the glass than on the intensity of the light. He decided to pass sunlight through a prism to split it into its component colours, and he used thermometers to monitor the heat of each colour. To his surprise he found that the heating effect was greatest at the red end of the spectrum, and that it continued beyond the visible light:

> ... the full red falls still short of the maximum of heat; which perhaps lies even a little beyond visible refraction. In this case, radiant heat will at least partly, if not chiefly, consist, if I may be permitted the expression, of invisible light.[14]

Joseph Banks, William's ally on so many occasions, was delighted at the discovery of what we know as infra-red rays. The implications, he rightly thought, far transcended those of Uranus: "I hope you will not be affronted when I tell you that highly as I prized the discovery of a new planet, I consider the separation of heat from light as a discovery pregnant with more important additions to science."[15] He would like to come to Slough to discuss the subject, and to bring with him distinguished company: Henry Cavendish, and the American, Benjamin Thompson (Count Rumford), who "says that your discovery is the most important since Sir I. Newton's death".

William's other notable discovery from this period came when the long programme of sweeps with the 20-ft began to draw to a close, and he felt free to re-examine some of the double stars he had identified a quarter of a century ago, to find out what had happened to them in the meantime. He found that in several pairs the two stars had altered their positions relative to each other, in ways that suggested they were companion stars, presumably in elliptical orbit about their common centre of gravity (though more observations would be needed to confirm this). John Michell had been right all along: an attractive force (or forces) operated among the stars, as Newton had shown it to do within the solar system.

William continued to make telescopes for sale, even though marriage had brought him affluence. The finest, and the most successful he ever made,[16] was commissioned by the King of Spain in January 1796. It was to be 25-ft in focal length, and of 24-inches aperture, bigger on both counts than William's large

20-ft, but still manageable. William completed it in 1798, although he did not ship it to Spain until 1802. Little if anything was achieved with it, for the mounting was destroyed by Napoleonic troops in 1808; but the two mirrors survive in Madrid to this day, along with the assembly instructions in Caroline's hand, illustrated by beautiful watercolours.

For his own use, and as an insurance against future infirmity, William made himself a reflector of just 10-ft focal length but with 24-inch mirrors. A mirror of this size had to be hollowed out more than usual if it was to bring the image to a focus in so short a distance; but William's skills were now equal to the challenge. The telescope could be directed to a given star or nebula in minutes and so was especially useful on nights that were partly clouded, or when he was entertaining visitors. Eventually, in 1816, he disposed of it to Lucien Bonaparte, brother of the Emperor Napoleon.[17]

One of the reflectors that William made for sale had no astronomical purpose, but was a contribution to the war effort against Britain's traditional enemy, France. We know of it only from a couple of letters Charles Burney wrote to William in September 1799, and an entry that month in Burney's *Memoirs*. William Pitt the younger was then Prime Minister, and George Canning his under-secretary for foreign affairs. Pitt evidently decided that an early-warning system against a French invasion was needed, and that a telescope by William installed at Walmer Castle overlooking the Straights of Dover was the answer. He therefore commissioned one at a cost of 100 guineas.

Herschel's telescope arrived at the Castle in kit form with a dozen or so pages of assembly instructions in Caroline's hand. As any DIY enthusiast knows, assembly instructions are all very well, but they assume a certain practical *nous* on the part of the user, and Canning and his friends were mere politicians and civil servants. "... no one had sufficient skill to put it together, and render it useful in examining the French coast and intermediate sea, for which it was originally intended."[18] Burney joined the group, to find the nation's early-warning system spread across the room in pieces.

A perplexed Canning turned to Burney for help; after all, he was the author of an astronomical poem, so he must know something about astronomy. Canning accordingly put Caroline's pages into Burney's hands, and stepped back expectantly, the problem solved. Burney, a poet and no astronomer, was dismayed. But rescue was at hand:

> ... before I had read six lines company poured in, and I replaced it [the

instructions] in the drawer whence Canning had taken it; and to say truth, without much reluctance, for I doubted my competence.[19]

Before anyone could think to call on him again, Burney sent an appeal to William for help. Was there someone in the Walmer area who understood how to assemble telescopes? If not, would William please send fuller instructions.

A couple of weeks later the Prime Minister himself was at Walmer, puzzling over the assembly of the telescope, with just two of the six steps completed:

> Mr Pitt perused your instructions with satisfaction; & when he came to the 6 short rules laid down — to the 2 first he said:— "we are right thus far — but the putting on the small speculum has puzzled us" — the rest he seemed to think not difficult. And said that he had left a person at work, who seemed to understand the business.[20]

What became of this telescope we do not know.[21]

Unfortunately, William devoted too much of the first decade of the new century to indulging an obsession with optical experiments on coloured rings; Newton's explanation of this phenomenon, he thought, did not hold water. The issue lay on the very margins of astronomy, but such was William's prestige that he had almost automatic right of publication in the pages of *Philosophical Transactions* for anything he cared to write. Friends repeatedly urged caution upon him, but to no avail, and his reputation suffered; after his death, when John was planning the republication of his father's collected papers, he decided that these articles were not worthy of inclusion.[22]

Caroline would be called upon to write out a fair copy of each paper destined for *Philosophical Transactions*, and on occasion this was no simple task — "My last paper consisted of eighty pages, so that you will have a piece of work to gather it together out of the scraps I leave"[23] — but otherwise her involvement in much of William's current work was marginal. Her appointment as salaried assistant to William had coincided with a reduction in his need for assistance, as the thrust of his work changed; henceforth it was less concerned with the acquisition of new information, more with reflections on information already to hand.

She had however identified one especially important item of deskwork for which she was uniquely suited, for it involved reorganizing large quantities of data; and when calculating or copying numbers she was never ever known to err, until a very occasional slip began to creep in when she was in her seventies.[24] During the early sweeps, William had been in awe of the reputation of the first Astronomer Royal, John Flamsteed, and of his great British Catalogue of nearly three thousand stars, the bible for students of the starry heavens.[25] From time to time William and Caroline came across a star in the sky that Flamsteed could scarcely have overlooked, yet it was not in the Catalogue; at other times a star that was in the Catalogue proved to be missing from the sky. "Taking it for granted that this catalogue was faultless", William later wrote, "I supposed them to be lost",[26] and he theorized over the fate that had befallen the stars that had disappeared.

However, it finally dawned on him that the problem lay in the Catalogue itself: it was not infallible, but contained errors and omissions. In particular, a listed star might be missing from the sky, not because it had moved away or diminished in brightness, but because its coordinates in the Catalogue had been wrongly transcribed from Flamsteed's original observations, or wrongly typeset by the printer. Yet there was no way of referring back from an entry in the Catalogue (in vol. iii of the *Historia Coelestis*) to the observations on which the Catalogue was based (in vol. ii), and therefore no way of tracking possible errors back to their source.

In 1795 William "recommended" Caroline to use her increased leisure to rectify this limitation in the Catalogue, by compiling an "Index to Flamsteed's observations of the fixed stars contained in the second volume of the *Historia Coelestis*". Her index — which took her the better part of two years to compile[27] — was divided into constellations, and within each constellation the stars were listed by their Flamsteed number; against each star was given (by printed page and line) the Flamsteed observations on which the catalogue entry was based.

Caroline also listed the brightness magnitudes recorded in Flamsteed's observations of a given star. The Catalogue itself gave only a single, average magnitude for the star, whereas each one of the original observations would have a magnitude; if these differed, this might be simply because of changes in viewing conditions or other such cause, but it might be because the star was a 'variable' that fluctuated in brightness and was therefore of special interest.

While making this collation, Caroline identified and listed no fewer than 561 stars that had been observed by Flamsteed but then overlooked in the compilation of the Catalogue: the Catalogue should have contained, not three thousand stars, but three-and-a-half thousand. She also listed the errata that she noticed.

As ever, her intention had been to facilitate William's observations.[28] But the Astronomer Royal recognized that her labours had turned Flamsteed's great work into one in which observers could have full confidence. He therefore persuaded the Royal Society itself to pay for the publication in book form of her three lists: omitted stars, index, and errata. In some copies the height of the page is a staggering 44cm, and the work has a title to match:

Catalogue of Stars, taken from Mr. Flamsteed's Observations contained in the second volume of the Historia Coelestis, *and not inserted in the British Catalogue, with an Index to point out every observation in that volume belonging to the stars of the British Catalogue. To which is added, a collection of errata that should be noticed in the same volume.* By Carolina Herschel. With introductory and explanatory remarks to each of them. By William Herschel, LL.D. F.R.S.

William is here at his most patronising: "And I may add, that my inspecting the work as it proceeded, and looking over all cases which seemed to require more of the habits of an astronomer than she has been in the way of acquiring, I have endeavoured, as much as I could, to prevent errors from finding their way into the work."[29]

What Caroline had done was immensely useful, yet it had required nothing more than perseverance and an infinite capacity for taking pains. In the world of astronomers and their assistants, only Caroline possessed these qualities.

Caroline was naturally delighted that her work was considered of such importance that the Royal Society had published it at their own expense. In thanking Maskelyne for his part in this, she admitted that this

has flattered my vanity not a little. You see, Sir, I do own myself to be vain because I would not wish to be singular, and was there ever a woman without vanity? — or a man either? only with this difference, that among gentlemen the commodity is generally stiled ambition.[30]

For all his support for the publication of Caroline's volume, Maskelyne harboured regrets that the position of each of her 561 omitted stars was specified relative to a nearby star; as a result no actual coordinates were given for the omitted stars, and this reduced the book's usefulness to him as an observer. Caroline lost no time in supplying him with what he wanted. She took up-to-date positions for the reference stars from a newly-published catalogue by the Rev. Francis Wollaston, who was a respected amateur; calculated the coordinates of the omitted stars; arranged them in the order in which they cross the meridian; and presented Maskelyne with the list in the form he wanted.[31] The fair copy she kept for herself ran to 25 folio pages of numbers. No wonder Caroline was held in such high regard among astronomers.

Her appetite whetted by the success of her Flamsteed catalogue, Caroline set herself another task. William's catalogues of nebulae were the work of a natural historian of the heavens. William had gone out at night in company with Caroline, to collect specimens. When he found one, he examined it, classified it, and gave it

The Herschel Partnership

A Catalogue of the Stars which have been observed by Wm Herschel in a series of Sweeps; brought into Zones of N.P. Distance and order of R.A. for the year 1800, by applying the variation given with each respective star in Wollaston's or Bode's Catalogues.

Circumpolar Stars from 0 to 10 Degrees

R.A in Time	N.P. Dist	Mag	Names, Characters, Observers &c	Sweeps	Date	G. No
3 36 41,5	9 54 4	6	36 Rangiferi 13° Cat. 49 Cephei Hev	763.	Oct 10.1787	·
8 40 18,0	8 23 44	8	184 Camelopardali 13° Cat. L	1111.	Sept 26 1802	·
9 7 19,6	7 48 46	4.5	186 P L	1111.	– – –	·
9 22 46,5	9 57 47	7	188 L	1111.	– – –	·
10 14 31,8	8 29 1	6.7	191 L	1111.	– – –	·
10 22 54,2	8 32 26	6.7	193 L	1111.	– – –	·
14 40 57	9 32	6		1074.	Dec 20,1797	1069
16 17 8	5 47	8		1106.	Jan 1, 1802	1107
16 27 15	6 3	8		1106.	– – –	1108
16 50 24	5 52	7		1106.	– – –	1109
17 5 52,6	7 40 21	4	ε Ursæ minoris .... H.	1106.	– – –	
17 34 56,7	9 42 13	5	4 Cephei 13° Cat. L	1075.	Dec 20, 1791	1071

Zone 10 Degrees N.P. Distance

R.A in Time	N.P. Dist	Mag	Names, Characters, Observers &c	Sweeps	Date	G. No
9 22 7,6	10 58 19	7.8	Red. 187 Camelopardali 13° L	1111.	Sept 26 1802	·
9 33 25	10 36 ·	8		1096.	Ap 2. 1801	1096
9 55 42	10 11 ·	8		1097.	Ap 10.1801	1099
13 42 23	10 51 ·	6		1074.	Dec 20,1797	1067
13 53 8,7	10 0 44	7	Two more. 220 Camelop. 13° L	1074.	– – –	1068
16 37 11,2	10 37 33	6+5.6	1 Cephei of 13° L	1075.1104	– – –	1070

Zone 11 Degrees N.P. Distance

R.A in Time	N.P. Dist	Mag	Names, Characters, Observers &c	Sweeps	Date	G. No
2 40 10,2	11 23 32	7.6	25 i Rangiferi 13°. 47 Cephei Hev	757.	Sept 16, 1787	·
10 43 52	11 1 ·	7	196 Camelopardali 13° .. L	1097	Ap 10 1801	·
11 54 53,8	58 38	6+6,5+7	206 L	1066.1068.1074	Dec 10,1797	1055
12 2 23	24 ·	7+7.5		1096.1099	Ap 2. 1801	1098
12 2 39,3	16 24	5+6+6,7	208 N Comelop. 13° 4 Drac. Hev	1074.1096.1099	Dec 20,1797	
13 12 54	55 ·	7		1065.	Dec 9, 97	1053
13 25 25,8	19 24	6+6,5	215 Camelopardali 13° .. L	1074.	Dec 20, 97	1066
14 10 4,5	31 17	5+5+5	4 b Ursa min. British Cat Cor.	1066.1069.1074	Dec 10, 97	✳
14 58 6,3	1 16	8	37 Ursa min. 13° L	1064.	Dec 6,1801	·
15 51 35,9	36 52	·5+4.5	16 ξ Ursa min. .... H	1061.1073.1104	Dec 10, 97	·
16 45 37	57 ·	7.8	The last & most north of 3.	1061.	– – –	1052
23 58 21,0	24 0	7+7	316 Cephei of 13° L	757.758	Sept 16, 87	783

Zone 12 Degrees N.P. Distance

R.A in Time	N.P. Dist	Mag	Names, Characters, Observers &c	Sweeps	Date	G. No
10 24 41	12 49 ·	6.7		1096	Ap 2,1801	1097
13 30 47	52 ·	7		1066	Dec 10,1797	1057
13 36 29	37 ·	6+7		1064.1065	Nov 22, 97	1050
13 38 37	9 ·	6+6	With 2 small stars preceding	1066. 1074	Dec 10. 97	1058
13 43 25	21 ·	6	With 2 S.ft. prec. A coarse treble	1064.	Nov 22. 97	1051

The first page of the catalogue Caroline prepared between June 1799 and March 1804, of the coordinates (for 1800) of the stars used to specify the positions of nebulae encountered in sweeps. When in Hanover in the early 1820s she was to use this catalogue to prepare a matching one specifying the coordinates of the nebulae themselves. The arrangement into 'zones' made these catalogues convenient to use when sweeping, and the 1820s volume was crucial in John's revision of his father's lists of nebulae. RAS MS C.3/2.3.

a serial number; he noted where it was to be found, and what it looked like. Thus the opening entry of his second catalogue informs us that on 28 April 1785, he came across his 94th bright nebula; a reader wishing to see it for himself should go to the star 61 Ursae, and proceed as directed (along the lines of "left a bit, up a bit").

This was satisfactory for an astronomer wishing to study samples of bright nebulae; but for an observer who chanced upon a nebula in the course of his night's work and who wished to identify it in William's catalogues, such a format was well-nigh useless. A first step would be to assign coordinates to each of the stars she had recorded in the course of the sweeping, and then to arrange the stars in the format she had used so successfully when compiling her list of Flamsteed stars for use in sweeps: by 'zones' of stars of roughly the same distance from the North Celestial Pole, the stars within each zone being ordered in the sequence in which they passed overhead.

Although the stars are so far away that for most purposes they still justify their Greek epithet of 'fixed', the Earth's axis has a pronounced wobble. As a result, the North Celestial Pole is constantly moving, and so the distance of any star from the Pole is constantly changing, as is its other coordinate. Because of this, any astronomical catalogue cites coordinates for a specific date or 'epoch'. Wollaston's catalogue gave star positions as they were in 1790; Caroline preferred to take the turn of the century for her epoch, even though this would involve additional calculations. In June 1799 she "Began re-calculating all the sweeps as a constant work for leisure time",[32] and on 16 March 1804, by which time she also possessed a copy of a new catalogue by J. E. Bode, "Finished re-calculating sweeps. Above 8,760 observations have been brought to 1800".[33] Her splendid manuscript volume, "A Catalogue of the Stars which have been observed by Wm Herschel in a series of Sweeps; brought into Zones of N[orth] P[olar] Distance and order of R.A. [the celestial equivalent of longitude] for the year 1800, by applying the variation given with each respective star in Wollaston's or Bode's Catalogues",[34] was to lie fallow for a quarter of a century. Then it would bear abundant fruit.

After William's marriage, with the slackening of the pace of their joint work, Caroline found herself with more time to sweep for comets. She discovered her second on 21 December 1788. Maskelyne was deeply impressed: "As it came up from the south it seems that Miss Herschel lost no time in finding it, I mean that it could not have been seen much sooner even in her excellent telescope."[35] But Caroline was not the only observer on the lookout for comets, and it transpired that Charles Messier in Paris had anticipated her.

She found two more comets in 1790, on 7 January and 18 April. As one discovery followed another, male astronomers home and abroad were captivated. The Astronomer Royal asked William, "Pray make my hearty acknowledgements to Miss Herschel for her meritorious attention to our science".[36] To Wollaston she is "Miss Herschel whom I put first as a sister astronomer".[37] Professor Karl Seyffer of Göttingen, who had been one of innumerable visitors to Slough, acclaims Caroline as "most noble and worthy priestess of the new heavens".[38] To M.-A. Pictet, professor at the Academy of Geneva, she is "the celebrated Miss Caroline".[39] Jerôme de Lalande of the Collège Royale in Paris addressed his letter of thanks for news of the fourth comet to "Mademoiselle Carolina Herschel, astronome célèbre, à Slough".[40] When writing to William, Lalande never failed to enquire after "*la savante* miss";[41] in September 1788, he sent "a thousand tender respects to *la savante* miss, of whom I frequently speak with enthusiasm".[42]

As we have seen, William himself was impressed enough by Caroline's early discoveries of comets to make her a bigger and better comet sweeper.[43] This was something of a mixed blessing, for the tube of the new instrument was over 5-ft long and the eyepiece was at the upper end — and Caroline was very short. On 6 August 1793 she writes: "Finding myself unable of bearing the fatigue of standing, I took the small sweeper which was in good order and shewed objects very well."[44] Her complaints continued: on 28 October "I met with many impediments. The telescope is raised too high and some steps are wanted."[45] Sometimes she used one sweeper and sometimes the other; and on at least one occasion, when they were trying to recover a comet Caroline had discovered some days before, William used the large sweeper while his sister used the small.[46]

It was with the large sweeper that Caroline found her fifth comet on 15 December 1791. Lalande now felt justified in nominating her for one of the prizes the Assemblée Nationale had recently established for "the most useful work or the most important discovery for the sciences or the arts". Unfortunately for Caroline, William's claim to the first prize was indisputable — Lalande had years before addressed a letter to "Monsieur Herschel, le plus célèbre astronome de l'univers, Windsor"[47] — and in the end it was not thought right to give two of the prizes to members of the same family.[48] Lalande's admiration for the two Herschels was such that after an evening spent with them stargazing, he wrote that "I have told everybody that I have never ever passed such an agreeable evening, not excepting even nights spent in love".[49]

Just how much time Caroline was currently spending in comet-sweeping is difficult to tell, for she normally drew a blank, and often she saw little purpose in

Within the image: *Catch*

What a strong sulphurous scent proceeds from this meteor

The Female Philosopher, smelling out the Comet

Pub. Feb[?] 1790 by R Hawkins N°53 at Truman st S[?] Soho

A cartoon dated February 1790 and showing "The Female Philosopher" — undoubtedly Caroline — "smelling out the Comet" (although her words put into her mouth would imply that she did not know the difference between a comet and a meteor). Caroline found a comet on 7 January of that year. Peel Collection, Pierpont Morgan Library, New York, vol. 9, p. 77v, no. 245.

recording this. But her account of the night of 14 March 1793 is unusually complete, and is worth reproducing in full as it demonstrates the extraordinary dedication of which she was capable:

> I looked at the Moon to see if any luminous spots were visible in the dark part [William was convinced he had seen evidence of lunar volcanoes] but saw none. Meanwhile I was looking, which was about half past 7, an occultation of a small star took place.
>
> I swept for comets from the head of Taurus and left of[f] with γ in the foot of Andromeda. The rest of the night was spent [as amanuensis to William at the telescope] in writing down observations in the Miscelaneous Journal till 1 o'clock. Then I began to sweep with β Cassiopeiae perpendicular sweeps from the horizon and upwards of 50°. I left off a[t] 4ʰ common time [4 a.m.] with γ Cygni & κ Pegasi.
>
> From 4 till 4ʰ30′ I swept with horizontal sweeps near the horizon that part of the heavens which was in the beginning of this sweep below the horizon. I saw nothing except the 52ᵈ of the Connoissance des Temps [that is, M 52] and a nebulous patch between β & κ Cassiopeiae which I believe is my faint nebula Obs. Sept 27. 1783.[50]

Caroline found her sixth comet on 7 October 1793 (though again she had been forestalled by Messier), and her seventh on 7 November 1795. This was later to become famous in astronomy as Encke's Comet. In 1818 Johann F. Encke of Seeberg Observatory, near Gotha, succeeded in computing the orbit of a comet that had recently been observed, and he found that it returns to Earth every 3.3 years, being identical to the one seen by Caroline in 1795, and indeed, in 1786, by Méchain before her.[51] Caroline may well have learned of this; but if so, there is nothing to suggest that it aroused her interest.

Her last comet discovery occurred on 14 August 1797. William was away at the time, and

> at 9ʰ 30′ being dark enough for sweeping I began in the usual manner with looking over the heavens with the naked eye, and immediately saw a comet.... I went down from the observatory to call my brother Alexander that he might assist me at the clock. In my way in the Garden I was met & detained by Lord [Storker?] and another Gentleman who came to see my brother and his telescopes. By way of preventing too long an interruption I told the Gentlemen that I had just found a comet and wanted to settle its place. I pointed it out to them and after having seen it they took their leave.[52]

William being absent, it fell to Caroline to communicate her discovery to professional astronomers. No time was to be lost; the letter to Maskelyne announcing her second comet had taken the best part of 48 hours to travel from Slough to Greenwich "owing to the slowness of our penny post",[53] and by then the weather had turned bad with the result that Maskelyne had not been able to observe the object for several nights. Such a delay was unacceptable and must not happen this time. As she wrote to Sir Joseph Banks on 17 August:

> This is not a letter from an astronomer to the President of the Royal Society announcing a comet, but only a few lines from Caroline Herschel to a friend of her brother's, by way of apology for not sending intelligence of that kind immediately where they are due.
>
> I have so little faith in the expedition of messengers of all descriptions that I undertook to be my own, with an intention of stopping in town and write and deliver a letter [to you] myself, but unfortunately I undertook the task with only the preparation of one hour's sleep, and having in the course of five years never rode above two miles at a time, the twenty to London, and the idea of six or seven more to Greenwich in reserve, totally unfitted me for any action. Dr. Maskelyne was so kind as to take some pains to persuade me to go this morning to pay my respects to Sir Joseph, but I thought a woman who knows so little of the world ought not to aim at such an honour, but go home, where she ought to be, as soon as possible.[54]

This valiant lady, after a night's observing followed by just one hour's sleep, saddled a horse and rode the best part of thirty miles to Greenwich so that the Astronomer Royal should be able to view the comet that same evening.

Within weeks of this discovery, as we shall see, Caroline made the unfortunate decision to move into lodgings. Thereafter, if she undertook an observing session, she would face the problem of how to make her way home in the dark. Not surprisingly, her career as a comet-seeker went into decline. After the turn of the century her memoranda of observing sessions become increasingly sporadic, and many concern comets whose existence had already been announced. After the entries for 1813 there is just one for 1815, one for 1817, one for 1819, and a final, poignant one for 1824. By then, Caroline's days as an observer in her own right were long gone.

Caroline won international renown for her successes in detecting the arrival of comets, and rightly so. But we look in vain in her manuscripts for any hint of interest beyond the mere act of discovery: in the orbits of her comets and in the possibility of previous and future apparitions, to say nothing of the physical nature of comets, and their role in the economy of nature (a topic of the greatest interest to William[55]). Once she had found a comet she proudly handed it over to astronomers, her job done. Her brother Dietrich expressed her attitude perfectly, even if Caroline rejected what he said: "... whenever he [Dietrich] catches a fly with a leg more than usual, he says it is as good as catching a comet!"[56]

Meanwhile, late in 1789, the great telescope had at last been brought to completion. William had made trial observations with the first mirror as early as February 1787, but it was too thin and did not maintain its shape when tilted in the tube. As it weighed nearly half a ton, its polishing presented problems very different from those posed by the small mirrors of his Bath days, when Caroline would feed him by spoon while he worked with the mirror in his hands. Now the polishing was done by two teams each of a dozen men,[57] with numbers on their backs so that William could marshal their movements.[58]

> The garden and workrooms were swarming with labourers and workmen, smiths and carpenters going to and fro between the forge and the forty-foot machinery, and I ought not to forget that there is not one screw-bolt about the whole apparatus but what was fixed under the immediate eye of my brother. I have seen him lie stretched many an hour in a burning sun, across the top beam whilst the iron work for the various motions was being fixed.[59]

A second, thicker mirror was cast in February 1788;[60] it weighed nearly a ton, and now the men needed were sometimes as many as twenty-two, and even that number found it heavy going.[61] William therefore devised machinery for the purpose, and this greatly reduced the number of men he needed. On one occasion he even tried horse-power, but "the animal could not be managed to go fast or slow or to stop at command";[62] it was dismissed in disgrace after just half an hour.[63]

The reflector became one of the sights of Britain, lying as it did alongside the London to Bath road that passed through the High Street at Slough. In his *The Poet at the Breakfast Table* the American author Oliver Wendel Holmes described it as

> a mighty bewilderment of slanted masts, spars and ladders and ropes, from the midst of which a vast tube, looking as if it might be a piece of ordnance such as the revolted angels battered the walls of Heaven with, according to Milton, lifted its mighty muzzle defiantly towards the sky.[64]

It was sufficiently prominent to be marked on the Ordnance Survey map (see page 58),[65] and popular magazines compared it to the Colossus of Rhodes and the

Porcelain Tower of Nankin.[66]

But the monster was barely completed when it lost what was surely its primary purpose. If — *if* — all nebulae were simply clusters of stars disguised by distance, then the 40-ft would surely help confirm this, by resolving into their component stars a number of prominent (and no doubt relatively near) nebulae that had resisted resolution by the 20-ft. But in 1790, while sweeping with the 20-ft, William came across what he (rightly) satisfied himself was indeed an example of 'true nebulosity' — a nebula that was *not* a distant star cluster.[67] As far as William was concerned the issue was now settled — some nebulae were star clusters, others were not — and so any contribution to the debate from the 40-ft would be of minor interest. On the other hand, too much money, effort and prestige had been invested in the monster for it to be relegated to the sidelines.

Moreover, the great and the good began to beat an incessant path to William's door. To the Royal Family an evening outing to Slough provided a solution to the problem of how to entertain visitors to Windsor Castle. The Queen was a patron of music, and her band, which included four, and later five, grandsons of Isaac Herschel, would play while the guests were at dinner.[68] The King was a patron of astronomy, who could claim credit for the greatest of telescopes, built and staffed at his expense by a son and daughter of Isaac; and in the evening royal guests would be taken to visit the observatory and see the famous telescope for themselves.

The pride aroused in the mind of William's royal patron by the thought of his great reflector was matched by the expectations of the scientific community. The professor of astronomy at Cracow could not leave for home without visiting "le premier sanctuaire de cette science".[69] The head of the Paris Observatory wrote that he was crossing the Channel primarily to see William and his telescopes.[70] The professor at the Collège Royale planned to travel to Slough "lui rendre mes hommages, et voir votre superbe telescope de 40 pieds".[71] The 40-ft dwarfed all other telescopes in existence, and envious observers worldwide expected from it one revelation after another.

They waited in vain. William's notebook "Observations with the 40 feet Telescope"[72] is admittedly not a complete record; but the pages are mostly blank, observations being recorded on a total of just seventeen days, between October 1788 and April 1793. Almost the only items of interest are the discovery of two additions to the five satellites of Saturn that were previously known.

This planet had always held a particular fascination for William, and when he had opened his first Observing Book back in March 1774, Saturn had been the subject of the very first entry.[73] On 28 August 1789 he turned the newly-completed

40-ft towards it.

> Having brought the telescope to the parallel of Saturn, I began to sweep with
> it.... 23[h] 30. Saturn with five stars in a line. The nearest of the five is probably
> a satellite which has hitherto escaped observation. It is less bright than the
> others. What makes me take it immediately for a satellite is its exactly ranging
> with the other 4, and with the ring.[74]

On 17 September he could go further: "I see six satellites at once, and being
perfectly assured that the 2[d] is invisible it becomes evident that Saturn has 7
satellites."[75]

No one had done more than Banks to facilitate the building of the 40-ft, and
William had been anxious to show him that his support had not been misplaced.
The letter William wrote him announcing the discovery of the sixth satellite was
followed within days by another announcing the seventh:

> Perhaps I ought to make an apology for troubling you again with a letter on
> the same subject as my former one; but if satellites will come in the way of my
> 40 feet Reflector, it is a little hard to resist discovering them.[76]

The euphoric William, in his fanfare in *Philosophical Transactions* announcing to
the world the arrival on the scene of the 40-ft, stores up trouble to come.

> But it will be seen presently, from the situation and size of the satellites, that
> we could hardly expect to discover them till a telescope of the dimensions
> and aperture of my forty-feet reflector should be constructed; and I need not
> observe how much we Members of this [Royal] Society must feel ourselves
> obliged to our Royal Patron, for his encouragement of the sciences, when we
> perceive that the discovery of these satellites is intirely owing to the liberal
> support whereby our most benevolent King has enabled his humble astronomer
> to complete the arduous undertaking of constructing this instrument.[77]

Doubtless such discoveries were no more than Banks had expected, and he and
the astronomical community worldwide sat back in the confident expectation of
a stream of further revelations.

They waited in vain. For the rest of its career, the great telescope was to achieve
next to nothing. Even its claim to the discovery of the two satellites of Saturn was
questionable. William had previously seen the first in August 1787 when observing
with the 20-ft, but had been too busy at the time to follow up the observation;
and he had seen the second with the 20-ft on 8 September 1789 — its claimed
'discovery' with the 40-ft a few days later had been little more than confirmation.

He left his published account of the discoveries ambiguous on such points, and his reader is often under the impression that he is reporting observations with the 40-ft, when in fact they should be credited to the 20-ft.[78]

In the years ahead William had to do what little he could to fend off the criticisms of disappointed colleagues. To sweep the whole sky with the 40-ft, he explained in 1799, would take many centuries.[79] Opportunities to use it, he wrote in 1814, were "very scarce", if only because the mirror would be covered with condensation or ice for hours, days or even weeks.[80] "The preparations for observing with it take up much time", he explained in 1815:

> The temperature of the air for observations that must not be interrupted, is often too changeable to use an instrument that will not easily accommodate itself to the change: and since this telescope, besides the assistant at the clock and writing desk, requires moreover the attendance of two workmen to execute the necessary movements, it cannot be convenient to have every thing prepared for occasional lucid intervals between flying clouds that may chance to occur; whereas in less than ten minutes, the twenty feet telescope may be properly adjusted and directed....[81]

Before long astronomers came to realize that the 40-ft was a failure, but the outings from Windsor Castle never abated. In 1818, three decades after the monster saw first light, and with its constructor in his eightieth year, visitors to William's observatory included: Princess Elizabeth and the Prince of Hesse Homburg with a count and two barons (April); the Prince and Princess Shaumburg von der Lippe (June); the Archduke Michael of Russia "with numerous attendance" (July); Princess Sophia of Gloucester, the Archbishop of Canterbury and several Lords and Ladies (August); and the Ertz Herzog Maximilian of Austria (October).[82] William had little option but to struggle, decade after decade, to conceal from his royal patrons the truth about the great telescope, and to maintain at least the semblance of a functioning instrument.

The one serviceable mirror needed repolishing every second year if possible, certainly every third or fourth year.[83] William was sometimes able to arrange for this to be done outside the Bath season so that Alexander would be free to come and help him. But as the brothers grew older, they found it an increasing strain to supervise the work and carry it through to a successful conclusion, for manhandling the one-ton mirror could be dangerous. Caroline describes an incident that occurred on 22 September 1807:

In taking the forty-foot mirror out of the tube, the beam to which the tackle is fixed broke in the middle, but fortunately not before it was nearly lowered into its carriage, &c., &c. Both my brothers had a narrow escape of being crushed to death.[84]

Nevertheless, if Caroline is to be believed,[85] repolishing continued until the summer of 1814, when William was 75.[86] Caroline noted: "His strength is now, and has for the last two or three years not been equal to the labour required for polishing forty-foot mirrors."[87] William last used the reflector (on Saturn) in August 1815, even though "the mirror is extremely tarnished".[88]

Yet still he would not give up. If the mirror was no longer functional, this would not be immediately evident to the outside world, for he allowed no-one else to use the telescope; it was the mighty structure that his visitors came to admire (if they were allowed to view the heavens it was with lesser instruments). And so, three years later, in July 1818, with his eightieth birthday approaching, William attempted to restore the mounting; "but the great heat he was exposed to by directing the workmen who are repairing the woodwork, &c., &c., of the 40 ft is surely too much for him".[89]

The telescope had become a millstone around William's neck. Eventually in private he had to admit defeat: "The woodwork is fast decaying and cannot be effectually mended, and ... I cannot recommend the 40 feet to be kept up."[90] Yet pride would not allow him simply to abandon the instrument and make public admission of failure. For this Caroline paid a price:

> But when all hopes for the return of vigour and strength necessary for resuming the unfinished task [of polishing] was gone, all cheerfulness and spirits had also forsaken him, and his temper was changed from the sweetest almost to a pettish one; and for that reason I was obliged to refrain from troubling him with any questions, though ever so necessary, for fear of irritating or fatiguing him.[91]

Because Caroline destroyed her records of the decade after William's marriage, we are poorly informed about her activities in this period other than her cometary discoveries; but it seems that for some years she continued to live in the cottage accommodation as had been agreed between William and Mary. Before long the Slough house was enlarged, probably in 1794. The laying of the corner stone was an occasion that stuck in Caroline's memory because the two-year-old John was invited to place the customary coin on the stone, but the sensible infant proved reluctant to surrender the money in such a futile cause.[92] The construction suggests that the arrangement between William and Mary for dual residence came to an end about this time, and that it was then that they established the Slough house as their single home.

For some wholly unexplained reason, in October 1797 Caroline made the disastrous decision to give up her convenient quarters in the cottage adjacent to the house of her brother and sister-in-law, and went to lodge and board with one of William's workmen.[93] There is no hint in any surviving document of the slightest rift between Caroline and Mary, and a rift between Caroline and William was unthinkable; but for some reason she left. When John's wife questioned her in her old age about her memories of John's childhood, she excused the limitations of her reply by explaining, in diplomatic language, that when he was five, "I came to be detached from the family circle".[94] It would not be the last time that Caroline acted in haste and repented at leisure.

In July of the following year, while William was on holiday with his family, Caroline "went daily to the Observatory and work-rooms to work, and returned home to my meals, and at night, ... in fine weather, I spent some hours on the roof, and was fetched home" by the workman with whom she lodged.[95] Just how inconvenient were these new arrangements is illustrated by a memorandum from the end of 1798: "Uncommonly harassed in consequence of the loss of time necessary for going backward and forward, and not having immediate access to each book or paper at the moment when wanted."[96]

Around the turn of the century her domestic arrangements were rarely settled for more than a few months at a time. 15 July 1799: "Agreed for apartments at Newby's, the tailor, in Slough."[97] 19 November 1799: "The bailiffs took possession of my landlord's goods, and I found my property was not safe in my new habitation."[98] 29 June 1800, in preparation for some weeks she was to spend in Bath, "Everything was arranged for my books and furniture to remain at my lodging, to which my brother was to keep the keys. But on receiving information they would be seized along with my landlord's goods by bailiffs, I prepared the same night for their removal, and all was safely lodged in a garret at [Mary Herschel's] by July 2 at night".[99] Returning from Bath towards the end of the year, she took rooms at Windsor with her nephew George Griesbach,[100] only to move four months later to the village of Chalvy where she rented a small house from a woodcutter.[101] It was all very disruptive and unsatisfactory.

After two years at Chalvy,[102] she arranged to live in rooms at Mary's former home at Upton, and at last some stability came to her domestic arrangements, for she would be there for seven years. The property had a common boundary with that now occupied by William and Mary, although there was half a mile between the two houses: an easy walk across the fields for Caroline in daylight, if the skies were clear and William intended to observe that evening, but more difficult in the dark. If overcast skies unexpectedly cleared during the night, there was (or so it is said) a local boy on whose window she would rap; he would come down with a lantern, ready to comply with her request, "Please will you take me to my Broder".[103]

In August 1799 Caroline had spent several days with the Astronomer Royal at Greenwich. She took with her William's copy of the second volume of Flamsteed's catalogue, so as to transcribe her marginal comments into Dr Maskelyne's copy. But the entertainment laid on for her was lavish, and she made the promised transcription only by working through the night.[104]

Periods such as these, spent away from home and her routine work, were exceptional. Caroline's life for the most part consisted in making herself useful to William in whatever way she could, going from her lodgings to the house over which she had once presided, and working there at her desk. When her surviving memoranda resume once more, towards the end of the century, the entries are mostly brief, and often this is simply because there was little of note to record. The heroic days were past, and William's astronomical researches would soon consist more of discussion of the implications of his vast store of past observations than of efforts to add to the collection. As William moved towards old age, life for his astronomical assistant became routine and even humdrum.

The year 1800 is not untypical. The final catalogue of 500 nebulae and clusters was not yet complete, yet — astonishingly — not a single sweep is recorded for that year. On 19 February William began his re-examination of double stars. He had just returned from over a fortnight at his holiday home in Bath,[105] where the season was in full swing; on 28 April he was off again, this time for two weeks in London. At the end of July he went with his family for a holiday on the south coast that lasted for the whole of August and into September. In late November he was off again to Bath for three weeks. William was now into his early sixties, the nights he had spent at the telescope were taking their toll, and Mary had persuaded him to slow down and take advantage of the prosperity they enjoyed through her three inheritances and his flourishing trade in telescopes. These repeated holidays were a far cry from the fanatical dedication of his early days in astronomy: the cry of "Let me but get at it again!" was no longer to be heard.

A stern critic might even wonder if King George was getting value for the grants he had awarded. A surviving account dated 26 July 1800, said to be in the King's

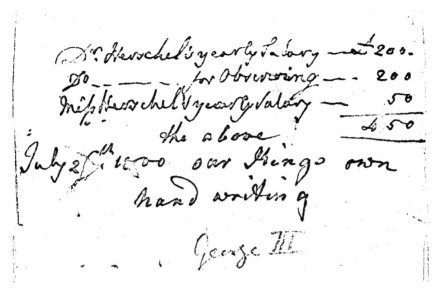

A note (in the King's own hand, according to Mary) for the salaries due to William and Caroline in 1800, together with the agreed running expenses. Document in private possession.

own hand, lists not only the £200 for William and the £50 for Caroline, but also the £200 for observing expenses with the 40ft. Yet a search through the observing records suggests that between February 1798 and April 1801, the 40-ft was used on only two nights;[106] if so, William was claiming expenses that he had not in fact incurred.

William's absences on holiday left his salaried assistant with the problem of how to employ her time. On 28 April 1800, following William's departure for London, "I was at the Observatory after he was gone, from ten till two, to select work for me to do at home".[107] Next day, "From ten till three at the Observatory to make order in the books and MSS". Early May, "Worked every day some hours at the Observatory".

William had had a house in Bath for some time; not only was Alexander living there, but William had many friends in the town, among them William Watson,[108] who was shortly to be mayor,[109] and Edward Pigott, who had moved there from York. By the summer of 1800 William had given up his previous house there and taken a different one, and he wanted Caroline to make it habitable. On 25 June she started to pack her things for an extended stay in Bath, "for there I am to go!"[110] Do we detect irritation at being ordered around by one of her brothers without so much as a 'please'?

On her arrival at Bath, Alexander met her at William's new holiday home. It had been empty before William took it over, but workmen had brought the furniture from William's previous house, and all was in chaos. The orderly Caroline made an inventory before hiring "a maid of all work to assist me to bring the house into habitable order, and by July 29th I was ready for resuming the work of recalculating sweeps, or despatching some copying, &c., which was sent me by the coach from Slough, and from the printer in London".[111]

On 10 September, with the Bath season not yet under way, Alexander left her for two weeks to go to Slough and help William repolish the mirror of the 40-ft. "Some of my time during his absence I spent at his house on Margaret's Hill to clean and repair his furniture, and making his habitation comfortable against his return."[112]

At the end of October, "I received notice that in about a fortnight I should be wanted at Slough".[113] Before he set off again William planned to show her the work that needed doing — and, he added, but almost as an afterthought, "I may have a few sweeps". On 24 November he departed for Bath, leaving Caroline with the keys to the Observatory "to make order and for despatching memorandums". But before the year ended, her brother had returned, "extremely ill".[114]

The years that followed saw a steady decline in William's health, and Caroline's memoranda present a melancholy record of one illness after another. 24 January 1807: "My Brother returned with a violent cough, added to a nervous headache...."[115] 26 February 1808: "My Brother so ill that I was not allowed to see him, and till March 8 his life was depaired of, and by Mar. 10th I was permitted to see him, but only for two or three minutes, for he is not allowed to speak."[116] 9 May 1808: "My Brother returned, nearly recovered, but with a violent cold and cough caught on the journey."[117]

Caroline's own health was far from good, if we are to believe her own account. In 1814 she records: "I, for my part, felt I should never be anything else but an invalid for life, but which I very carefully kept to myself, as I wished to be useful to my brother as long as I possibly could."[118] Allowance must be made for her customary pessimism: she belongs to the group of people who invariably have influenza, thinking the common cold beneath them. But no one could dispute that in 1808 she had an unpleasant scare. On 2 October "I was very ill and had Dr. Pope to attend me". Payments to Pope recur repeatedly in Mary Herschel's accounts,[119] but Caroline was less easily satisfied. A week later, "dismissed Pope and went to Dr. Phips".[120] Phips, it seems, decided she was going blind, and that she should use the days of sight remaining to her to practise for life under this handicap. She was therefore kept in a darkened room for a fortnight.[121] But the alarm was without foundation: "Nov. 20. Phips pronounced me out of danger for becoming blind, which he ought to have done much sooner, or rather not to have put me unnecessarily under such dreadful apprehensions."[122]

An unexpected joy for Caroline came when she was reunited with a long-forgotten acquaintance with whom she had shared lessons in millinery and dressmaking when they were girls together in Hanover. Caroline had been seventeen at the time, and among the other pupils had been "a sweet little girl about ten or eleven years old".[123] In 1802, at Queen's Lodge within the precincts of Windsor Castle, Caroline was introduced to a married lady, Mme Beckedorff, who had a position in the Queen's household. To Caroline's astonishment, Mme Beckedorff greeted her warmly, as her fellow-pupil from Hanover. The two became firm friends and confidants, and it was through Mme Beckedorff that Caroline got to know many members of the Royal Family.

> But I was soon sensible of having found what hitherto I had looked for in vain — a sincere and uninterested [disinterested] Friend, to whom I might have applied for counsel and comfort in my deserted situation.[124]

In March 1806 the quiet routine of her life was given a welcome interruption when William received a letter from Dietrich saying he hoped to visit England later that year. His own son had recently died in America, their brother Jacob was unmarried when he met his violent death in 1792, Alexander was a widower without children, and so William's son John was the only one left to carry on the Herschel name. "I long to see good old Alexander, and Caroline who is so kind to us; and to see your son would alone be worth the journey; *he* will soon be the only one of our race."[125] Dietrich duly arrived on 6 July, and spent most of the month at Slough. He left on the 24th; and the normal rhythm of life resumed.

Caroline is usually supremely indifferent to what was going on in the world outside her own little circle, but for her native Hanover she made an exception. In 1803, as we have seen, she had accepted from Mary a quarterly addition of £10 to her salary from the King because she had "been in a panic for my friends at Hanover".[126] The country had become a pawn in the power struggle between Napoleon and the King of Prussia, and in July 1808, "We received very distressing accounts from our brother in Hanover".[127] Things went from bad to worse, and eventually Dietrich decided that his only hope of supporting his family was to leave them, go to England, ply his trade as a musician, and send money home. He arrived at Slough early in November.

From the hour of Dietrich's arrival in England till that of his departure, which
was not till nearly four years after, I had not a day's respite from accumulated
trouble and anxiety, for he came ruined in health, spirit, and fortune, and,
according to the old Hanoverian custom, I was the only one from whom all
domestic comforts were expected. I hope I have acquitted myself to everybody's
satisfaction, for I never neglected my eldest [surviving] brother's business, and
the time I bestowed on Dietrich was taken entirely from my sleep or from what
is generally allowed for meals, which were mostly taken running, or sometimes
forgotten entirely.[128]

Dietrich, living the life of a bachelor during his unhappy exile from Hanover, needed
his sister; Caroline, it must be said, needed to be needed. Indeed, with the decline
in William's health, her thoughts and affections were turning more and more
towards Dietrich, nearly seventeen years William's junior and still on the right side
of sixty. He was currently in desperate need of money, and so throughout his four
years in England Caroline made over to him the £10 a quarter she was receiving
from Mary.[129]

Eventually, in the summer of 1812, Dietrich decided to return to Hanover. He
would have to go by a roundabout route, though this would involve a "precarious"
journey through Sweden. Time passed, and at Slough his brother and sister waited
anxiously for news. At the end of July they had word that he had got safely as far
as Göttingen, "but of receiving any further account we had not the least prospect,
for all communication, with Hanover in particular, was cut off".[130] A letter Dietrich wrote on 15 August failed to reach Slough; but as it happened, he had been
robbed of his pocketbook during the journey, and so had to write to England for
a duplicate of one of the documents stolen from him. Such letters were allowed
to pass through France, and the eventual recipient was able to reassure William
and Caroline that Dietrich was indeed alive and well. In due course his situation
there improved, and in 1816 Caroline notes the arrival of "very satisfactory letters
from our brother in Hanover".[131]

Caroline missed him. While he had been with them she had regularly seen all
three of her surviving brothers. Dietrich in particular was a person to whom she
felt able to turn. Later, when worrying over the disposal of her few effects, she was
to remark: "my whole life almost has passed away in the delusion that next to my
eldest [surviving] brother, none but Dietrich was capable of giving me advice."[132]
Indeed Dietrich was in some ways even closer to her than William, for Caroline
says that in their later years she "kept to the resolution of never opening my lips

to my dear brother William about worldly or serious concerns".[133]

The next to return to Hanover was Alexander. He had lived in Bath since 1770, as bachelor, husband, and widower. He had long been William's closest ally in the physical labour of building telescopes, he had taken part in the abortive attempts in 1781 to cast 3-ft mirrors in the basement of their house, he had assisted in their move to Datchet, and he had since spent many summers with them while Bath was out of season, helping make telescopes for sale or polishing the mirror of the 40-ft. Early in 1816, William and Caroline had "a most melancholy letter"[134] from him to say that he had injured his knee and was confined to bed. The injury failed to heal, and this was a serious matter for one who was essentially an itinerant musician. He had been taking more pupils than ever and had hoped to clear his remaining debts,[135] but now he was on crutches. Caroline offered to go and care for him, and although he declined, it soon became clear that he could no longer manage alone.

Alexander was now approaching seventy, and eventually it was agreed that the time had come for him to retire to Hanover. Caroline accompanied him to Wapping,[136] and there the captain promised to see him safely to Bremen where a friend of Dietrich's awaited him. Alexander's repatriation proved a success, and he lived contentedly in Hanover for the rest of his life.

Caroline next lost to Hanover her closest friend outside her family.[137] After the death of Queen Charlotte in November 1818, Mme Beckedorff too returned to her native town, since her services at Court were no longer needed. As Caroline wrote a few months before William's own death, "... but now when most I want her, she is gone!"[138] Gradually, Caroline's circle of family and friends in England was diminishing, and her circle in Hanover increasing.

William apart, the principal counter-attraction to Hanover lay in her nephew John, "the young prodigy"[139] who would one day be buried in Westminster Abbey alongside Isaac Newton. John entered Cambridge in 1809 at the age of seventeen, and in 1813 graduated as 'Senior Wrangler', first among the first-class honours in a brilliant year. Within weeks he was elected a Fellow of the Royal Society, and in 1821, when he was still in his twenties, he was awarded the Society's Copley Medal, the very honour that his father had received for the discovery of Uranus. In 1816, William, in his late seventies and at the culmination of a long and distinguished career, was made a knight of the Guelphic Order, and became Sir William; John received the same honour in 1831, when he was not yet forty.

Caroline's residence at Mary Herschel's former home had lasted until 1810, when it was arranged that she should move closer to William. It is often said that she occupied the house on Windsor Road between William's home and the Crown Inn where Mary's mother had lived, and which now belonged to William.[140] But it seems likely that she took the Annexe to the Inn, around the corner on the High Street,[141] for Caroline speaks of it as "attached to the Crown Inn",[142] and William calls it "the House at the corner".[143] Caroline did all the redecorating herself,[144] William taking her to Windsor to choose the paper for the sitting room.[145] But although she was living a mere two hundred yards from the home of William and Mary, she was lonely: the three last months of 1812, for example, she spent "mostly in solitude at home, except when I was wanted to assist my brother at night or in his library".[146]

Caroline lived at the house for four years. William's gardener's wife, Mrs Cock, cleaned and cooked for her, in addition to tending William's hens for which she received a guinea a year.[147] In 1814 Mr Cock fell ill, and his wife could not leave his bedside to attend to Caroline. Worse, it became clear that Cock would never work again,

> and providing for him in a different manner for his support, I became unsettled once more, as the house we occupied was to be let, and I was obliged to move to a small cottage in Slough, at a considerable distance from my Brother.[148]

It seems likely that the Inn required use of the Annexe, perhaps in order to extend its accommodation.[149]

The move led to a modification in Caroline's pattern of life, for visits to William now required planning. When Mary was absent, Caroline would go over and spend the nights at William's; when Mary returned, Caroline would do her "calculating and copying" in her own cottage.[150]

Meanwhile, despite William's declining health, the succession of papers sent to *Philosophical Transactions* continued. Between 1811 and 1818 he published — among other things — four major studies explaining 'the construction of the heavens' as he now understood it, drawing on his vast experience as a natural historian of the heavens. Gravity was the agent of change. A thinly diffused cloud of nebulosity would begin to condense here and there, where the nebulosity was more concentrated than elsewhere. Out of these concentrations stars would be born, and in time the scattered stars would be drawn together into clusters. These would become ever more concentrated, until in the aftermath of a cataclysmic gravitational collapse the whole process would begin all over again.

William had, he said, seen light that had been two million years on its journey to his telescope; in such contexts, "millions of years are but moments".[151] The grandeur of his conception is breathtaking, as he encourages astronomers to lift their eyes beyond their parochial preoccupation with our own restricted region of the universe, and reveals the riches awaiting their attention in the stellar cosmos. In lieu of the endlessly repeating cycles of the clockwork universe of Newton, he lays out nebulae and clusters for inspection in order of age, selecting from his vast catalogues first the young, then those in maturity, and finally the elderly, comparing what he offers to "an annual description of the human figure, were it given from the birth of a child till he comes to be a man in his prime".[152] The words were William's; but the handwriting was always Caroline's.

William at this stage in his life was providing little new evidence, but rather reflections on his incomparable treasury of observations. The importance of preserving these past records was coming to dominate his thoughts. As it happened, sorting books and papers had for some time kept Caroline occupied when, as so often, there was little else for her to do. So, for example, she writes in 1806:

> I distracted my thoughts by undertaking an amazing deal of work; among the rest, I made catalogues of all books and MSS. my brother's library contained, and arranged them, to the best of my knowledge, according to what the confined room would allow.[153]

By 1819 the preservation of his records had begun to prey on William's mind. So, when he was about to set off for a holiday in Bath,

> The last moments before he stept in the carriage were spent in walking with me through his Library and workrooms, pointing with anxious looks to every shelf and drawer, desiring me to examine all, and to make memorandums of them as well as I could. He was hardly able to support himself and his spirits so low, that I found difficulty in commanding my voice so far as to give him the ashurance he should find on his return that my time had not been misspent.
>
> When I was left alone I found on looking around that I had no easy task to perform, for there were packs of writings to be examined which had not been looked at for the last 40 years. But I did not lose a single day without working in the Library as long as I could read a letter without candle-light, and taking with me copying, &c. &c., which employed me for best part of the night, and thus I was enabled to give my Brother a clear account of what had been done at his return on the 1st of May. But he came home worse than he went, and for several days hardly noticed my handiworks.[154]

As long as William lived, he fretted over the safety of his papers:

... that the anxious care for his papers and workrooms never ended but with his life was proved by his frequent wispering enquiries if they were locked up and the key safe; of which I took care to assure him that they were, and the key in Lady Herschel's hands.[155]

Lacking a photocopier, William even had Caroline transcribe in pen and ink his first dozen papers in *Philosophical Transactions*,[156] which he needed to complete a set for binding.[157] She, as so often, claims to have achieved this only by going without sleep.

There was so much that William had been unable to do — most vexing of all, he had never even glimpsed the southern skies, and he never would. The only hope of completing the Herschelian exploration of the construction of the heavens lay with John.

In 1814 John had left Cambridge and embarked on a career in law, but his heart was not in it. Such was his scientific versatility that the following year he applied for the Cambridge Chair of Chemistry, and lost by only a single vote.[158] That vote was to have implications for the history of astronomy, and perhaps for the history of chemistry; but the failure would not prevent him becoming one of the leading pioneers of photography.

Despite this disappointment John soon returned to his old college, St John's, to what was no more than a modest position teaching mathematics, although within months, in 1816, he achieved appropriate status as a Fellow. That spring William had cunningly begun to whet John's interests in telescopes, by having him use the 10-ft destined for Lucien Bonaparte to read a card in a tree nearly a thousand feet away.[159] That summer they went on holiday together to Dawlish on the Devon coast. When the right moment came, William pressed John to abandon the career as a Cambridge don on which he had set his heart, and to return home to become his father's apprentice.

William was very persuasive. In October John wrote to Charles Babbage:

I always used to abuse Cambridge as you will know with very little mercy or measure, but, upon my soul, now I am about to leave it, my heart dies within me. I am going, under my father's directions, to take up the series of his observations where he has left them (for he has now pretty well given over regularly observing) and continuing his scrutiny of the heavens with powerful telescopes.[160]

Theories could be learned from books and periodicals; the craft of telescope

1816. September 17

I made a sweep with the 20 ft telescope and took the transit of some stars — among which were 9β Capricorni & 13 Capricorni the apparent times of their passage compared with the times of meridian passage deduced from Wollaston's catalogue ~~seemed to indicate~~ a deviation of the telescope from the meridian to the East, amounting to 16'14" or 16'16" of time, and an error in the quadrant of 45' or 41' See the annexed memorandum.

Sept. 18.

This deviation was corrected by moving the telescope in azimuth through an angle determined by a proper calculation. The rubbing plate too having been loosened & deranged, my Father & my-self fastened it, and adjusted it by a plumbline to a vertical position — We likewise corrected the quadrant by the adjusting screws

At night I viewed (7 feet) ζ Ursæ — γ Andromedæ — ζ Aquarii — the Clusters in the Swordhandle of Perseus — 15th H.b. of Connois- des T. — 11th d° d° — &c. having previously adjusted the mirrors & eyeglasses in the day time, and adjusted the finder by a star.

Sept. 30. 1816.

Micrometers. The distance of a post in the hedge from the outward corner of the door-frame of my Father's book-room was found to measure 2452·85 inches. —

The 10 feet tel. directed to an object or post, was placed so as to have vertex of obj. Mirror 7·75ᵗʰ from the corner of the door frame — leaving 2445·1ᵗʰ = dist from post

The focal length on object was then taken = 125·44"

Hence to find focus on the stars

D = Distance of object = 2445·1
F = focal length on obj = 125·44
f = focal length on stars.

$$f = \frac{F \cdot D}{F + D} = 119·32''$$

The object viewed was a 6 inch Ruler divided into 10ths. Nairne's Micrometer of 50 feet, parallel wires, 30 measures of different lengths upon the ruler were taken of which a statement is annexed. — (NB. The measures were all taken closing, and by interior contacts.)

A new star rises in the firmament of astronomical observers: the first entries in John's Journal no. 1. John would one day be buried alongside Newton in Westminster Abbey. RAS MS J.1/10, 1.

Rough notes, mostly in Caroline's hand and partly in pencil, of the two nights in May 1821 when William and Caroline launched John on his career of 'sweeping' for nebulae. Caroline's pencil records the comments of James South, who was present, on seeing a nebula with two nuclei: "O! good God! It is worth going to the devil for!"

making could be learned only through an apprenticeship. The 40-ft was beyond redemption, but the 20-ft was not:

> The woodwork of it is also decaying, but the whole apparatus may be renewed at a moderate expense, and the method of repolishing and preserving or restoring the figure of the mirror when tarnished is not attended with such critical difficulties as will occur when the weight of the mirror requires a crane to lift it on and off the polisher.[161]

The telescope — or so it seemed — needed no more than an input of youthful vigour under the supervision of a trained hand. In 1817 two new mirrors were cast, and John polished one of them under William's direction, finishing it in December of the following year. The other mirror he polished himself. John believed he now had the instrument with which to finish his father's work.

In December 1820 John received a visit from his friend, the wealthy amateur astronomer James South, and together they turned the 20-ft on the Moon and Saturn. John was mightily impressed: "I can now believe anything of the effect of reflectors of great aperture." Then disaster struck. The telescope

> took it into its head to fly from its center, and I fear will require to be taken to pieces before it can be put right again. I am not sorry for this; it will afford my poor Father some occupation, which (though able to do very little) he stands much in need of, and is quite a new man when superintending some little repairs, &c.[162]

A month later John was giving directions "to a pack of carpenters for erecting a *new 20-feet Telescope*. The old one is completely off its legs".[163] John later regarded the "new" 20-ft as the last of his father's telescopes,[164] "and the clearness and precision of his directions during its execution, showed a mind unbroken by age, and still capable of turning all the resources of former experience to the best account".[165]

John now had to learn how to sweep. On 1 June 1821 the eighty-two-year-old William, summoning up all his reserves of strength, helped place one of the 18-inch mirrors in the tube, and sweeping began.[166] He himself was much too infirm to climb up and join John on the platform. Even Caroline was in her seventies. Yet she was anxious to give John the best possible start to his career as a Herschel. In years to come, John's records of sweeps were to prove as voluminous as his father's. The records of the first two sweeps, however, are not in his hand, but in that of Caroline.[167] After an interval of two decades, Caroline was back at her desk, a partner once more in sweeps for nebulae.

William could now, like Simeon on seeing the baby Jesus, utter his *nunc dimittis*. In October Caroline "closed my Day-book, for one day passed like another, except that I, from my daily calls, returned to my solitary and cheerless home with increased anxiety for each following day".[168] She sought comfort in the autobiography she was writing for Dietrich.[169] Months passed, and William became ever more frail; but his obsession with the 40-ft remained with him to the end. On 15 August 1822, the grandson of the Bulmans with whom William had lodged in Leeds so long ago asked for some token he could send to his father,[170] and William sent Caroline to the library "to fetch one of his last papers and a plate of the forty-feet telescope".[171] Ten days later William was dead.

5

Hanoverian Twilight

"O! why did I leave England!"[1]

When William's "heartrending" struggle against increasing infirmity finally came to an end, Caroline found that "not one comfort was left to me but that of retiring to the chamber of death, there to ruminate without interruption on my isolated situation".[2] Half a century before, she had been rescued by William from Hanoverian servitude. But her boundless gratitude had not been enough to stop him marrying, and so robbing her of primacy in his affections and in his home; and now he lay dead. Yet her loyalty to him had survived the first blow, and it would survive the second: she would do all in her power to foster his memory, and to remind John of his "sacred duty"[3] to complete his father's work.

Alexander had died in Hanover in March of the previous year. But Caroline's native town was still home to Dietrich, once her delightful baby brother and now the only survivor of her nine brothers and sisters. Dietrich's branch of the family were by all accounts "noble-hearted and perfect beings",[4] and the two of them had cemented their childhood bond during the four years that Dietrich had spent as a refugee in England. By 1820 Caroline had already decided to make her home with him when William died:

> I sent in 1820 thirty pounds to be laid out for a fether bed for me when after a long dreaded melancholy event I should be obliged to seek consolation in the busom of Diterich's family, which after the description of his wife and daughters' characters, I thought to be the only place on earth where I could find rest.[5]

Why did she not remain in England? It was a question she was to ask herself over and over again in the years to come. William's widow had been her sister-in-law for thirty years and more, and Caroline's early resentment of her had been replaced by a tender affection, as the contentment that marriage had brought William became evident. Caroline was one day to write to John from Hanover: "Tell your dear mother she must not give me the slip, for I will and cannot mourn for anyone more that I love."[6] Yet they had never been close, and throughout Caroline's years in England one senses a gap between them that affection could not bridge. They were a little wary of each other, women who had in their different ways competed for the same man.

John of course had long been the apple of her eye, but he was forty years her junior, a young man with an elderly aunt. And despite his relative youth he was already a major figure of the British scientific establishment, a leading light of the Royal Society and one of the founders of the Astronomical Society of London. The focus of his life was not Slough but the exclusively-male scientific life of the metropolis, in which Caroline had no part to play. And even as his father's scientific heir, John had no immediate need of Caroline; he was currently engaged in the study, not of nebulae, but of William's other great observational pursuit, double stars. This work required precision measurements, and so called for instruments quite different from the 20-ft cosmological battering ram. Fortunately John's friend, James South, possessed the very instruments they needed; the two collaborated in the measurements, working independently, and afterwards comparing results.[7] There was no place in this for John's aunt.

Caroline thought at the time that she might not live another twelvemonth.[8] But in fact she had another quarter of a century before her; and when eventually John's energies focused on the revision and completion of his father's work on nebulae, work in which she had played a crucial part, she would many times regret her absence from his side.

But the die was cast: she had bought her bed, and it awaited her in Hanover. Not only that, but in August 1822, with William's life ebbing away, Caroline had further strengthened the ties that bound her to Dietrich. After receiving "very distressing accounts of family misfortunes from my brother at Hanover",[9] she had decided to make over to Dietrich her life's savings of £500. Her pension from the Crown had been confirmed and her needs were few.[10] She did however take the precaution of qualifying the gift, by reserving the income from the investment for herself.[11]

With William dead, the moment had come for her to return home, to Hanover and Dietrich. The effort this required was itself a solace in her grief.

> My aunt though greatly distressed has borne this affliction with uncommon
> fortitude. She has resolved on leaving England immediately and going to reside

with her family in Hanover, and the expectation of preparing for her journey has been of service in distracting her attention from dwelling on its cause.[12]

So insistent was she, so imperative was it that she depart at once, that John took time from arranging William's funeral and consoling his distraught mother, to write to Dietrich about her plans.

> My aunt's fixed determination to quit England before the winter will render it impossible for me to do what I had earnestly wished, viz. to see her [safely] in Hanover, as it will be impossible for me to arrange all my dear Father's affairs soon enough and I cannot yet leave my Mother who requires my presence and support. But I hope, indeed I have no doubt, she will be in safe hands, and I shall only request you to give me the earliest accounts of her arrival, as it will take some time before she herself can be settled enough to write.[13]

William was buried on 7 September. Just two days later Caroline had already begun to select the books and clothing she would take, and to sell or give away her furniture.[14]

Dietrich, to his credit, would not let Caroline undertake the Channel crossing alone, and he came to England to fetch her, arriving at Slough on 3 October. On the 7th Caroline took farewell of Princess Augusta and her other Windsor friends, and three days later she and Dietrich departed for London. Many of the scientific luminaries in whose company she had always felt at ease — Maskelyne, Banks, Aubert — were dead, but she did not lack for friends in high places: she spent the 14th with Princess Mathilda, who sent a carriage to fetch her.

On the 16th she went to Bedford Place in London, "where all my friends were assembled".

> From all these sorrowing friends and connections I was obliged to take an everlasting leave, and in the few hours we were for the last time together, I was obliged to sign many papers, among which was a receipt for a half year's legacy. I signed this with great reluctance ... but Lady H. and my nephew insisted on my taking it, according to my brother's will. This unexpected sum has enabled me to furnish myself with many conveniences on my arrival here, of which otherwise I should have perhaps debarred myself.[15]

The same day John wrote to Charles Babbage:

> My Aunt has now been a week in Town & her stay in England will be but a day or two longer. She sails for Rotterdam in her brother's company who came

over to fetch her, and is well & in good spirits. She has exerted herself and made all her arrangements with extraordinary vigour.[16]

The obligation to take "everlasting leave" was her own doing. The expensive Hanoverian feather bed she had purchased two years earlier, even the £500 she had recently made over to Dietrich, had involved no irrevocable commitment. But in anticipation of her bereavement she had mentally turned her back on Slough with its melancholy memories, and returned in spirit to her childhood roots. Now it was too late to alter course.

Next day Caroline and Dietrich went to an inn near the Tower of London, and dealt with the customs formalities. John "came for a moment to us, and after his departure I saw no one I knew or who cared for me".[17] The following morning they boarded the steam packet. In her life Caroline was to make just two sea voyages, separated by half a century. The first had been memorable; the second proved equally so. After a stormy passage lasting forty-eight hours they arrived at Rotterdam.

> At one time a spray conveyed a bucket-full of water into my bed, which was regarded as nothing in comparison to the evils with which I was surrounded. I was the most sick of all on board, and the poor old lady was pitied by all.... At Blackwall we lay still three hours, then we hobbled on to near Gravesend, and there lay in a high sea at anchor all night, whilst they were hatching and thumping to mend the vessel we were to go in [to be taken ashore]. At half past eleven I set foot on shore, where so many people were assembled to gaze on us that it set me a crying.[18]

No sooner was she landed than the awful truth dawned, that Dietrich would not be the soul-mate she yearned for.

> But in the last hope of finding in Dietrich a brother to whom I might communicate all my thoughts of past, present, and future, I saw myself disappointed the very first day of our travelling on land. For let me touch on what topic I would, he maintained the contrary, which I soon saw was done merely because he would allow no one to know anything but himself.[19]

She was especially sad when Dietrich "for ever murmured at having received too scanty an education, though he had the same schooling we all of us had before him".[20] She resented this criticism of their father, "that excellent being" whose memory she cherished and whose battered almanac would one day share her coffin.

A week later she was in her native Hanover. Caroline was to lodge in the same house as Dietrich's family, and things began to improve. Despite suffering from rheumatism, Dietrich's wife

> is still of so cheerful a disposition and so active by way of overcoming disease by exercise, that I cannot wonder enough, and her reception of me was truly gratifying; the handsomest rooms, three or four times larger than what I have been used to, from which I can step in her own apartments, have been prepared for me and furnished in the most elegant style.[21]

Perhaps Caroline had made the right decision after all.

But Hanover had changed. "It is quite a new world, peopled with new beings, to what I left it in 1772."[22]

> What little I have seen of Hanover ... I do not like! And though some streets have been enlarged (as I am told), they appear to me much less than I left them fifty years ago.[23]

There were compensations. Within two hours of Caroline's arrival in Hanover, her bosom friend Mme Beckedorff sent to ask after her.[24] Before long the proud grandmother was lining up her ten grandchildren and presenting them to "Great Aunt Caroline", to the mystification of the children who found the diminutive old lady far from great.[25] Mme Beckedorff was to die some years before Caroline,[26] but her daughter continued as Caroline's confidant, and it was she who would one day write to John with news of his aunt's death.[27]

By December Caroline's trunks from England had at last arrived and she had clothes to wear. Soon she became known for her regular attendance at the Theatre, where she was to have a season ticket for fifteen years.[28] But partying was alien to her nature, and she was beginning to feel the weight of years: "As yet I lead but a dull sort of life; the town is much too gay for me — plays, concerts, card parties, walking, &c." And Dietrich was far from well. Indeed it had been noble of him to come to England to fetch her, for he was just recovering from a serious illness. His daughter had recently become a widow at the age of 37, only a few days after the birth of her ninth child; and travelling back and forth to offer what help he could had been too much for his weak constitution.

Before long Caroline's conscience began to trouble her. Why had she not remained in England and supported John in his "sacred duty"? "Believe me", she wrote to him after only a few months there,

> I would not have gone without at least having made the offer of my service for some time longer to you, my dear nephew, had I not felt it would be in

vain to struggle any longer against age and infirmity.... I preferred [Hanover] to remaining where I should have had to bewail my inability of making myself useful any longer.[29]

She considered — perhaps correctly — that whatever help it was within her power to give could be given at her desk, and there were desks in Hanover.

John had all the financial resources necessary to complete his father's work, for in his will William had left John both property and £25,000, a vast sum.[30] But still Caroline wanted to contribute her mite, offering in 1824 to forgo the £100 she received annually under William's will if this would help[31] — though in truth her eagerness to further the sacred cause may have been a pretext for declining money that had a taint about it, so engrained was her childhood detestation of being financially beholden to any brother.

Caroline constantly fretted over the annuity, her misgivings growing with the passage of the years. In 1834 John wrote:

Respecting the £100 received from Galterman [an agent], this is not the first instance I have had of the difficulty of persuading my good old Aunt to retain that portion of her very moderate income which it is my office to remit to her. Hitherto with more or less difficulty I have succeeded, but with constantly increasing resistance, and on one occasion a sort of conflict of this kind took place which placed poor Galterman at his wits ends by contrary orders, and which was only settled by my personal appearance in Hanover where I *made* her pocket it. As the matter stands at present however I feel convinced that she will be much more mortified by the return of the sum in question than by [your] retaining it — so you must retain it in hand as assets on my account and I hope by earnest expostulation to prevent a similar occurrence next year.[32]

Her eagerness to help found a more creative outlet after John wrote in August 1823 to say that "I hope this season to commence a series of observations with the twenty-foot reflector, which is now in fine order".[33] John must be encouraged to undertake the major challenge of revising William's vast catalogues of nebulae and star clusters. But the present format of the catalogues made this impracticable. It was not feasible for John to manoeuvre his reflector towards one particular star, and examine the nearby nebula; and then turn the instrument to a completely different part of the sky and do the same; and then point it in yet another direction to find a third nebula. This would take forever. Instead, he would have to examine the heavens systematically, sweeping horizontal strips of sky as William had swept, setting the 20-ft to a given North Polar Distance and moving it a little up and down as the sky drifted past; and he had no Caroline at a nearby desk to help him. It was essential that he know in advance the nebulae he might expect to encounter in the next sweep, so that he could be ready to verify — and if necessary revise — the position and description given by William. For this he needed the position of each of William's 2,500 nebulae and clusters to be given in coordinates, with the entries organized into a format convenient for an observer engaged in sweeps.

This was where Caroline could help. At the turn of the century she had compiled a catalogue of the stars that featured in the sweeps, calculating their positions for the year 1800 and arranging them in a 'zone' format ideal for use in sweeping; and she had brought this volume with her to Hanover, along with the records of the sweeps themselves.[34] In William's catalogues, the position of each nebula was given by reference to one of these stars, and so it would a straightforward (though onerous) task to produce a catalogue of nebulae, in a format matching that of her catalogue of stars. She replied immediately: "... I wish to live a little longer, that I might make you a more correct catalogue of the 2,500 nebulae, which is not even begun, but hope to be able to make it my next winter's amusement."[35]

She began by going through the records of sweeps and numbering the nebulae serially, for future reference (see pages 74 and 75). This was by no means as simple as it sounds, for sweeps overlapped and many nebulae had been observed on more than one occasion. For each nebula she then noted its identification in the original publication; the date when it was first observed; the reference star for the nebula; and the position of the nebula in relation to the star. She then took her manuscript catalogue of stars for the epoch 1800, and from the coordinates of the star she calculated the coordinates of the nebula. Finally she noted the reference number of the sweep (or sweeps) in which the nebula was observed.

This done, she then grouped the nebulae in 'zones' of North Polar Distance, according to their angular distance from the North Pole. The first zone consisted of the most northerly nebulae, lying less than 10° from the Pole. The second included those between 10° and 15° from the Pole, the third those between 15° and 17°, after which the width of each zone was reduced to one degree only. Within each zone, the nebulae were ordered in the sequence in which they pass overhead.[36]

If, therefore, John planned to observe with the telescope at elevations corresponding to NPD of (say) between 30° and 31°, he could expect to encounter nebulae listed in the relevant zone of his aunt's catalogue, and in the order listed under that zone.

By the summer of 1824 the work was progressing well, but she was anxious that its final format should be exactly suited to John's needs. In September she heard from him that he would shortly be calling on her as he made his way home from one of his European tours. This was good, for in a few hours of conversation they would achieve what "would have otherwise caused us both a tedious and vexatious correspondence in the future".[37]

John's impression of life in the Herschel home in Hanover was altogether more favourable than the one Caroline customarily portrays. As ever, her bottle was half empty.

> I found her very comfortably situated in her brother's family, and with no cause to regret her change of country. They afforded me an excellent view of German domestic manners which are pleasing and peaceful.[38]

Following his visit all went smoothly with the great work, and by January 1825 the end was in sight: "I am now writing out the Catalogue of Nebulae, and am at zone 30°, and hope to finish it for the Easter messenger."[39] And so she did.[40] The folio volume, now in the Library of the Royal Society in London,[41] runs to no less than 104 pages of numbers. It was a magnificent, indeed crucial contribution to the revision of William's nebular legacy. True, the epoch was 1800 and the coordinate system had 'precessed' since then with the wobble of the Earth's axis; but the resulting changes were small, and John prepared tables of corrections that he could use when necessary.[42] John was later to write: "I learned fully to appreciate the skill, diligence, and accuracy which that indefatigable lady brought to bear on a task which only the most boundless devotion could have induced her to under-take, and enabled her to accomplish."[43] Sir David Brewster, the Scottish physicist, termed it "an extraordinary monument of the unextinguished ardour of a lady of seventy-five in the cause of abstract science".[44]

John's response to the arrival of Caroline's volume was immediate. He was at last in a position to embark on the revision of his father's catalogues, and this he would:

> I received this afternoon your most valuable packet containing your labours of the last year, which I shall prize, and more than prize — shall use myself, and make useful to others. A week ago I had the twenty-foot directed on the nebulae in Virgo, and determined afresh the right ascensions and polar dis-tances of thirty-six of them. These curious objects ... I shall now take into my especial charge — nobody else can see them.[45]

In 1827 he writes to tell Caroline about the progress of his sweeps:

> I find your Catalogue most useful. I always draw out from it a regular *working*

list for the night's sweep, and by that means have often been able to take as many as thirty or forty nebulae in a sweep.[46]

In acknowledging receipt of Caroline's volume, John was able to tell her the outcome of the measurements that he and South had made of double stars. Over forty years had passed since William had published his catalogues of doubles, and over twenty years since William had re-examined a handful of them, finding examples in which the components had orbited around each other. John and South had measures of 659 doubles. "Among these we have now verified not less than seventeen connected in binary systems in the way pointed out by my father."[47]

The reordering of the results of so many sweeps had reminded Caroline of past triumphs, and she wished she had had the vigour of her youth and so been able to assist John as she had William:

> I ... am only sorry that I cannot recall the health, eyesight, and *vigor* I was blessed with twenty or thirty years ago; for nothing else is wanting (and that is all) for my coming by the first steamboat to offer you the same assistance (when sweeping) as, by your father's instructions, I had been enabled to afford him....
>
> But above all, dear Nephew, I beg you will consider your health. Encroach not too much on the hours which should be given to sleep. I know how wretched and feverish one feels after two or three nights waking....[48]

Meanwhile Caroline was increasingly concerned about the well-being of Dietrich, whose companionship had been a prime motive for her return to Hanover. Caroline maintained peace between them by holding her tongue: "... to combat against infirmities and peevishness (the usual companions of old age) depends entirely on my exertion to bear the same without communication, for unfortunately we are never in the same mind."[49] Her letters speak constantly of his ill-health. In August 1826, for example, "his health is so very precarious, that I often think he will go before me".[50] Later in the year she remarks that "I hardly ever knew a man of his age labouring under more infirmities, nor bearing them with less patience than he does".[51] Perhaps Dietrich was deserving of more sympathy than this, for in January he died, leaving Caroline the sole survivor of her generation. She moved into an apartment of her own.[52]

About this time Caroline herself gave John instructions as to the eventual disposal of her own effects. She had brought to Hanover a 7-ft telescope, and "my sweeper",[53] in fact the smaller (and more easily portable) of the two that William had made her. But she soon found that they were of little use: "... at the heavens

The gold medal of the Astronomical Society of London (the future Royal Astronomical Society) presented to Caroline in 1828. Above Caroline's name is William's 40-ft reflector, the symbol of the Society, and over it is the Society's motto, "Whatever shines is to be noted down". The verso shows Isaac Newton and a phrase from the Latin poem by Edmond Halley published in the opening pages of Newton's *Principia*: "the cloud [of ignorance] dispelled by science". By permission of the Mistress and Fellows of Girton College, Cambridge, to whom the medal was presented in 1885 by Caroline's great-niece, Lady Gordon.

is no getting, for the high roofs of the opposite houses."[54] According to her Books of Observations, she made use of her telescopes on just one occasion, on 31 January 1824, when she sketched the position of a comet that had been visible for the past three weeks.[55]

If the 7-ft in question was one of the two mounted instruments of this size that William had in later life,[56] then Caroline had taken only the optics and had left the mounting behind, for she told John in August 1823:

> I am amusing myself with having the seven-foot mounted by Hohenbaum, though I have not even a prospect of a window for a whole constellation, but it shall stand in my room and be my monument.[57]

One might have expected her to have taken (or had shipped) William's portable, "skeleton" reflector, which had accompanied him on various tours and which is listed in an inventory made towards the end of his life: "The tube and stand consisting of bars to be screwed together and taken asunder so as to lie in a small compass for convenient carriage."[58] But examination of Caroline's instrument, now in the Science Museum in London, suggests it was in no way unusual, and so the exact source of the optics remains a mystery. Caroline was anxious to have in her room a reminder of the reflector with which William had discovered Uranus, and so she had the mounting made of painted deal.[59]

In February 1828 a signal honour was conferred on Caroline by the Astronomical Society of London (the future Royal Astronomical Society). Until recently the umbrella of the Royal Society had served the needs of the entire English scientific community, amateur and professional, but the development of individual sciences, such as astronomy, seemed to many to call for specialist societies to cater for specialist needs. Accordingly, in January 1820 a number of astronomers, including John, met and agreed to form what they decided to call The Astronomical Society. The Duke of Somerset, whose title conferred prestige on any organization of which he as a member, was invited to become President; John was Foreign Secretary, and William one of the Vice-Presidents. Banks, however, who had been at the helm of the Royal Society for over forty years, fiercely resented the competition, and prevailed upon Somerset to withdraw. William, whose debt to Banks was immense, and who was in any case too frail to take any active role, hesitated to accept the vacant post. But in June, Banks died. The invitation to William was renewed; and in February 1821 he duly became President, though he was never well enough to attend a meeting.[60]

The Astronomical Society had been quick to institute the award of medals, in imitation of the medals awarded by the Royal Society, and on 8 February 1828, it fell to John as the current President to make the customary speeches on the annual award of two such medals. He next called on James South to speak as Vice-President. South had been present on the historic evening in June 1821 when Caroline and John had swept together,[61] and he was an enthusiastic admirer of Caroline. He was happy to announce the Council's unanimous resolution:

> That a Gold Medal of this Society be given to Miss Caroline Herschel for her recent reduction, to January, 1800, of the Nebulae discovered by her illustrious brother, which may be considered as the completion of a series of exertions probably unparalleled either in magnitude or importance in the annals of astronomical labour.[62]

South it was who made the announcement because it was inappropriate for the President to speak in praise of his aunt.

Caroline was touched. She was already a celebrity in Hanover, and now friends were calling to congratulate her on the London award.[63] But in old age her devotion to her brother's memory was fast becoming an obsession. "I could say a great deal about the *clumsy speech* of the V.P. Whoever says *too much of me* says *too little of your father!*"[64]

Caroline was even considered for a Royal Medal of the Royal Society itself, but the lapse of time since the close of her active career in astronomy made her ineligible.[65]

Time was dragging on her hands: "I have the two dullest months before me, for the plays and concerts do not begin again till autumn; all families are either gone to the baths or to their villas, &c. My friends are all some dozen years younger than myself, and I cannot always, or but seldom, accept their invitations."[66] We possess a remarkably extensive correspondence between Caroline and John from these years, and they present a Caroline whose mental powers are unimpaired, but who is increasingly frail and increasingly regretful of her decision to leave England and return to Hanover. "I cannot help comparing the country [England] in which I have lived so long, with this in which I must end my days, and which is totally changed since I left it, and not one alive that I knew formerly, except my dear Mrs. Beckedorff."[67]

Regretful, and even resentful. In a letter to John in 1832 her bitterness knows no bounds. She had returned to Hanover to be with Dietrich and his family, and so the blame for her predicament must be laid at their door: "... pretty letters they can all write for by those I continually received when in England I have been so wohefully deceived."[68] How was she to know that Dietrich had altered so much? He could hardly be blamed for the changes brought about by his poor health, but did he have to brag of his legacy from William of £2000 and of his relationship to his famous brother?

... how dreadfully disappointed almost in the first few hours of our meeting after a lapse of 8 years I found such an alteration both in person & manners that I hardly knew the man again. For the havoc illness had made in his looks my heart was filled with pitty; but I could not but lament that his Brother's noble remembrance should have had so fatal an effect as to run away with his senses, so as to demand homage from everyone for being a capitalist ... and a brother of Wm Herschel.

On first arrival in Hanover Caroline had described her reception by Dietrich's wife as "truly gratifying", and her sister-in-law as "of so cheerful a disposition ... that I cannot wonder enough". But by now her bitterness over her current plight had persuaded Caroline that the woman who had welcomed her so warmly to Hanover a decade ago had been little short of a monster. She now remembered herself as having been "introduced to Madam H. a short corpulent woman upwards of 60, dressed like a girl of 20 without cap, her brown hair mixed with gray plated and the temples covered in huge artificial curls I almost shuddered back from her embrace".[69]

On reaching Hanover she had been delighted to find herself provided by Dietrich's wife with "the handsomest rooms, three or four times larger than what I have been used to, from which I can step in her own apartments, have been prepared for me and furnished in the most elegant style". Now, looking back through resentful eyes, her recollection was that Dietrich had

> handed me into my room which had been washed in the forenoon, no fire, there I dropt on a sopha my eyes fixed on the wet boards without a carpet which it was my first care to buy and one for Mme H. by way of giving no offence by showing a desire of having anything apart.[70]

Dietrich was dead, and his widow, a woman of "avarice", "never comes near me without she has some design on my purse". Her children are no better:

> But instead of experiencing the least attention from those who are in Hanover or from the absent Nieces when they are in Town I hardly come in for a hasty call just before they are on their way to dinner.

She concludes:

> ... of the events of the last 10 years I have spent here, I can only say that they have been a perfect tissue of disgusting vexations doubly painful to bear because I could not communicate my complaints to anyone; because they were against my immediate connections.[71]

Disposing of her effects was becoming an obsession. By 1830 her papers were already with John; unaware that she had nearly two more decades to live, she now sent him some of her astronomical books,[72] including the presentation copy to William of her *Catalogue of Stars, taken from Mr. Flamsteed's Observations*,[73] and the catalogue by Wollaston. But when preparing the box for despatch, she experienced "too great a reluctance at taking my leave of Flamsteed intirely", and so she sat down and copied out all the Flamsteed stars listed by Wollaston. The 94 folio pages of numbers[74] were of no use to anyone, other than their sentimental value to herself; but she simply could not resist copying numbers. This done, she wrote: "Adieu Flamsteed."[75]

One event that gave her immense pleasure was John's marriage, in March 1829. He would not be lonely as she had been, and still was.

John had visited her in 1824, and he visited her again in June 1832:

> I found my aunt wonderfully well and very nicely and comfortably lodged, and we have since been on the full trot. She runs about the town with me and skips up her two flights of stairs as light and fresh at least as *some folks* I could name who are not a fourth part of her age.... In the morning till eleven or twelve she is dull and weary; but as the day advances she gains life, and is quite 'fresh and funny' at ten or eleven p.m., and sings old rhymes, nay, even dances! to the great delight of all who see her.[76]

John must have discussed with his aunt the great expedition he had long been planning to the southern hemisphere, where skies were visible that had been forever below William's horizon. Only thus could John complete the Herschelian examination of the entire heavenly sphere. At first he had thought of going to Paramatta in Australia, but by now he had settled on the Cape of Good Hope.

Caroline's reaction to news of his plans had been that of the old warhorse who scents battle: "Ja! If I was thirty or forty years younger, and could go too? In Gottes Namen!"[77] But John had to tread carefully with her. Caroline asked him to take an early opportunity to examine a region that had fascinated his father: "As soon as your instrument is erected I wish you would see if there was not something remarkable in the lower part of the Scorpion to be found, for I remember your father returned several nights and years to the same spot, but could not satisfy himself about the uncommon appearance of that part of the heavens."[78] John did as instructed, and found the region full of beautiful clusters of stars.[79]

This was not the right answer. "It is not *clusters of stars* I want you to discover in the body of the Scorpion (or thereabout), for that does not answer my expectation, remembering having once heard your father, after a long awful silence, exclaim, 'Hier ist wahrhaftig ein Loch im Himmel !' ['Here indeed is a hole in the heavens!']"[80] (It had not occurred to William that dark obscuring matter might be blocking his view, and so he had supposed a total absence of visible objects in a given direction to indicate some sort of hole in the star system.) The diplomatic John lost no time in reassuring her that in Scorpio there were *both* clusters *and* spaces devoid of stars.[81]

As a relative was to write to John after her death, "She looked upon progress in science as so much detraction from her brother's fame, and even your investigations would have become a source of estrangement had she been with you."[82] In Caroline's eyes, John's duty was to confirm whatever William had suspected.

John, perhaps surprisingly, tended to share this attitude. For all the esteem in which he was universally held, there were those who complained that he took his "reverence" for his father's work "to excess", and that his attitude approached "idolatry".[83] After his father's death he and Caroline had destroyed most of William's preliminary drafts of papers, to make it seem that the papers sprang from his mind already in their final, perfected state (thereby frustrating later historians eager to understand the genesis of William's ideas).[84] More worryingly, in his highly influential textbooks, which went through innumerable editions, John propounded William's early model of the Galaxy as though it was current truth — with no warning to the reader that William himself had abandoned the model long ago, when he realized the falsity of the assumptions on which it was based.[85]

Not that even this parading of William's abandoned theories was likely to have satisfied Caroline. As far as she was concerned, no one would ever surpass her brother's exploits in any way. On 7 March 1841 John, with the best of intentions, sent her an article describing the achievements and plans of William Parsons, the future third Earl of Rosse.[86] Parsons had already built on his estate in Ireland a reflector with 3-ft mirrors, twice the diameter of John's 20-ft, and was well advanced with a "Leviathan" with 6-ft mirrors. World leadership in the building of great reflectors had at last passed out of the hands of the Herschels.

John was full of praise for Parsons: the Leviathan he would later describe as "an achievement of such magnitude ... that I want words to express my admiration for it".[87] Caroline took a different view. Parsons's "great telescope, which *shall* beat Sir William Herschel's all to nothing" — the idea was so absurd that she spent the rest of the day in laughter.[88]

John arrived at the Cape of Good Hope in January 1834. He had with him his 20-ft for use on objects where 'light-gathering power' was all-important, and a

precision instrument he had bought from South that was ideal for the measurement of double stars. He also took with him Caroline's large sweeper. Despite its
unsophisticated design, the sweeper would be ideal for a first reconnaissance, and
John could use it to get acquainted with these skies he had never seen before. One
wonders if Caroline herself had reminded him of how useful the sweeper would
be on first arrival.

In a campaign prodigious even by the standards of a Herschel, John extended
to the southern skies every aspect of William's work. Among much else he lists
1707 nebulae and star clusters, and 2102 double stars observed with the 20-ft and
a further 1081 observed with the instrument he had bought from South. Such was
the acclaim on his return for what he had achieved that the very next month he
was awarded the hereditary title of baronet, on the occasion of Queen Victoria's
coronation. John rarely observed again, and the 20-ft, with which he had scrutinized
the whole of the heavens both north and south, was allowed to rot in his basement.[89] A few months after his return he wrote to a friend: "… with the publication
of my South African observations (when it shall please God that shall happen) I
have made up my mind to consider my astronomical career as terminated."[90] In
the same letter he declined the professorship of astronomy at Oxford.

When John arrived back from the Cape, he had nobly fulfilled his promise to
his father — and to Caroline — to extend William's observational campaigns to
the southern skies. Within days he set off to visit his aunt, taking with him his
five-year-old son William James.[91] Caroline, now eighty-eight, was sure that the
lad would come to harm if she took her eyes off him for a moment. If she gave
him something to eat, it would poison him. "I rather suffered him to hunger than
would let him eat anything hurtful; indeed I would not let him eat anything at all
without his papa was present."[92] But his great-aunt had her good points too, as
an amused observer records: "Well! what do you say of such a person being able
to put her foot behind her back and scratch her ear, in imitation of a dog, with it,
in one of her merry moods."[93]

Both of them knew that this would be the last time they would meet, and
Caroline prepared a speech of final farewell. No doubt she had much to say. Her
brother could now rest in peace, his life's work — and hers — completed by John.
William lived on, not only in his son but in his grandson. But when the hour came,
she found that John and his son had already fled the town, unable to face up to
the pain of parting. Caroline was utterly devasted. She felt she could never go
through such a traumatic experience again, and in 1842 she wrote to John's wife
Margaret (whom she never met but for whom she had great affection): "I would

not wish on any account to see either my nephew, or you, my dear niece, again *in this world, for I could not bear the pain of parting once more*; but I trust I shall find and know you in the next."[94]

One subject that had long been worrying Caroline — and which she and John no doubt discussed — was the eventual fate of the two telescopes she had brought from England. They had long since changed from assets to liabilities, as it became clear that she would never again use them to observe, and she began to fret over what would happen to them when she died. In 1828 she had instructed John that the small sweeper was to go to Mme Beckedorff's daughter, who would remain Caroline's faithful friend for the twenty years left to her, and the 7-ft to a Dr Groskopff, "though I know it will only be a relic to him, but it will not be destroyed or sold for an old song".[95]

Three years later Miss Beckedorff was confirmed as the one who would "take my sweeper under her protection".[96] But after further consideration Caroline evidently decided that the 7-ft was too important to be given to a mere acquaintance, and in September 1837 she told her distinguished visitor Alexander von Humboldt that, as it was "a 7 foot telescope for which all the optical parts were made by her brother", John was to have it as a present on his homecoming from the Cape.[97] Next month she had the instrument professionally packed.[98]

John's visit to Hanover took place in July of the following year, and presumably the crated instrument was shipped to him soon thereafter. John would have appreciated the compliment, but he himself had no use for that or indeed any telescope; and so it sat in its unopened crate, John being unaware that Caroline had included in the shipment a number of astronomical books.[99] By now his family included no fewer than seven children, and the house at Slough was far too small for their needs. In the summer of 1839 John found a more suitable property near the village of Hawkhurst in Kent, and his furniture was transported there early in April 1840, John remaining behind to wind up his affairs at Slough.

He still had to decide what to do with the crate containing Caroline's telescope, and he thought the reflector might be a welcome gift for what was now the Royal Astronomical Society; William had been the first President, and the corporate seal of the Society featured the 40-ft. On the 14th he wrote to take soundings from an officer of the Society,[100] received a favourable response,[101] and by the 25th the reflector was in the Society's possession, presented jointly by Caroline and John.[102] One suspects that Caroline would have preferred it to remain with John, but at least it was "saved from being sold for an old song".[103]

On 24 October 1838,[104] no doubt having discussed the matter with John, she

Engraving by George Muller showing Caroline in Hanover at the age of 97. Miss Beckedorff (*Chr*, 375) had misgivings about this likeness: "I am sorry to say the drawing which I saw did not do justice to her intelligent countenance; the features are too strong, not feminine enough, and the expression too fierce."

gave her sweeper to a local friend, Ulrich Friedrich Hausmann.[105] Its optical parts were later presented by a descendent of his to the Historisches Museum in Hanover, where they now are.[106]

Caroline retained her mental faculties almost to the end, but she became increasingly peevish and irritable. With her salary and her pension she had all the money she needed; but she came to resent the financial help she had given and was still giving to Dietrich's family. On her arrival in Hanover she had found Dietrich's second daughter, Dorothea Richter, in such financial straights that Caroline had promised her niece should have the interest on the £500 Caroline had made over to Dietrich. When Caroline had given other help to members of Dietrich's family, every last gutengroschen had been recorded in the meticulous way she had learned from William on first arrival at Bath. Now, in 1840 she sat down and paintakingly calculated the total, over 8315 thalers if one included the 548 thalers "laid out for a plate where a Christmas present of the kind was expected".[107] She now knew just "how much I have thrown away on beings to whom I was under no obligation of *any* kind; and among whom there is not one who would sometimes sacrifice an hour to cheer me in a long winter's evening".[108]

In the same year she embarked on the writing of a second account of her early years. It was for John, whom his ninety-year-old aunt asks to "excuse the style and the spelling, &c, &c, on account of my eyesight being so uncertain"; but she assures him that "my memory is as good as ever", as indeed it was.[109] A year later, she sent him thirty-one pages of fair copy, written as ever in her firm and legible hand, telling of her life story to the day of her arrival in Bath in 1772. John's wife Margaret wrote to beg her to "*go on* with your memoir until you leave England and take up your residence in Hanover". She responded, and in August 1845, when she was ninety-five, John wrote to thank her for the latest instalment, which is again written in the clearest of hands and shows a memory undimmed with age. It ends with Dietrich's abortive break for freedom in 1777 and his arrival at Bath. John urged her to persevere — "If it is only at the rate of 3 or 4 lines a day it will be an amusing occupation.... You have no idea how it interests us and Marg^t is quite in raptures when she gets a few fresh pages of it"[110] — but her strength was at last beginning to fail.

Honours came her way. In 1835 the all-male Royal Astronomical Society elected her to honorary fellowship; three years later she was made a member of the Royal Irish Academy;[111] and in 1846 the King of Prussia awarded her the Gold Medal for Science, "In recognition of the valuable services rendered to Astronomy by you as the fellow-worker of your immortal brother, Sir William Herschel".[112]

The same year, her visitors included the formidable George Biddel Airy, the successor-but-one of Nevil Maskelyne as Astronomer Royal. As usual, her guest was astonished at the vigour and intelligence of a woman of ninety-six; as usual, Caroline was full of grumbles.

She was not up when we called at 11 but was up at 2. She complained of not being well, but seemed to us to be extremely well, and possessing powers of body and mind which I could never expect to see in a person of such an age. She spoke partly in German and partly in English, of course we lost a little, especially as her articulation is slightly defective, but generally we got on very well.

She made mention of a catalogue of Flamsteed's stars to which she had attached notes, and some papers put into Gauss's hands....[113]

On her ninety-seventh birthday she was visited by the Crown Prince and Princess, who brought her the gift of a velvet armchair; she entertained them for two hours, and sang them a catch that William had written.[114]

That summer John was at last able to send her the sumptuous volume of his *Cape Observations*. When it reaches you, he tells her in a separate letter, "You will then have in your hands the completion of my father's work".[115] A few months later, on 9 January 1848, she died, her "unquiet heart" at rest at last.[116] Her funeral took place on the 18th. By order of the Crown Princess, her coffin was adorned with palm branches, and as a mark of respect the King of Hanover, and the Crown Prince and Princess, sent their coaches to follow the hearse.[117] Inside her coffin, at her request, was placed a lock of her brother William's hair, and an almanac that had been used by her father.[118]

Caroline's grave in Hanover (photograph by Owen Gingerich).

6

Retrospect

I have perhaps deserved it ... by perseverance and
exertions beyond female strength! well done![1]

In speeches delivered at the award of medals, nothing but good is said of the
recipient, and the speaker feels under no obligation to display both sides of the
coin. James South, who made the address at the meeting of the Astronomical
Society when Caroline was awarded its Gold Medal, had seen her at work, and his
admiration for the unconquerable old lady was genuine and sincere. We can sense
his wonderment, shared by astronomers throughout Europe, at the sheer scale of
her contribution to William's epoch-making reorientation of astronomy.

> Who participated in his toils? Who braved with him the inclemency of the
> weather? Who shared his privations? A female. Who was she? His sister. Miss
> Herschel it was who by *night* acted as his amanuensis: she it was whose pen
> conveyed to paper his observations as they issued from his lips; she it was
> who noted the right ascensions and polar distances of the objects observed;
> she it was who, having passed the night near the instrument, took the rough
> manuscripts to her cottage at the dawn of the day and produced a fair copy of
> the night's work on the following morning; she it was who planned the labour
> of each succeeding night; she it was who reduced every observation, made
> every calculation; she it was who arranged everything in systematic order; and
> she it was who helped him to obtain his imperishable name.[2]

South next lists the comets she discovered, and mentions the nebulae she contributed to William's catalogues, commenting:

> Indeed, in looking at these joint labours of these extraordinary personages, we
> scarcely know whether most to admire the intellectual power of the brother,
> or the unconquerable industry of the sister.

Finally he comes to the first of the great contributions she made at her desk:

> In the year 1797 she presented to the Royal Society a Catalogue of 560 stars taken from Flamsteed's observations, and not inserted in the British Catalogue, together with a collection of errata that should be noticed in the same volume.

He concludes with mention of the manuscript volume she had compiled to facilitate John's revision of his father's nebulae, "thus bringing to a close half a century spent in astronomical labour".[3]

That this should have been achieved by a woman who in childhood had been denied a training in simple arithmetic is astonishing; and contemporary astronomers — without exception male — were the more impressed because it was the work of someone who was "very little, very gentle, very modest, and very ingenuous; and her manners are those of a person unhackneyed and unawed by the world, yet desirous to meet and return its smiles".[4]

Helping William had been Caroline's fulfilment in life. As her father had long ago predicted, she died an old maid, starved of affection. The reader of her memoirs and letters is struck by her repeated lament at the lack of someone who "cared for me". As a child, when her father and William were absent with the band, "there was no one who cared anything about me". Sixty-seven years later, after saying farewell to John in London before embarking for Hanover, "I saw no one I knew or who cared for me".

As with so many celibates, Caroline's lack of affection was made worse by the want of someone in whom to confide. William was to lose that role, if indeed he had ever held it: Caroline "kept to the resolution of never opening my lips to my dear brother William about worldly or serious concerns".[5] In 1802 Mme Beckedorff had miraculously appeared to fulfil this need, and Caroline "was soon sensible of having found what hitherto I had looked for in vain — a sincere and uninterested [disinterested] Friend, to whom I might have applied for counsel and comfort in my deserted situation."[6] But in 1822, when Caroline had to make a final decision as to whether or not to spend the rest of her life in England, Mme Beckedorff was back in Hanover: "... when most I want her, she is gone!"[7] Dietrich was the other potential confidant: "my whole life almost has passed away in the delusion that next to [William], none but Dietrich was capable of giving me advice."[8] But Dietrich too was in Hanover when the decision had to be made. And after the decision was irreversible, he was revealed as a man of straw.

Life dealt Caroline a poor hand. Her childhood and adolescence had been a time of unremitting labour, undertaken against a background of war. From her earliest years she had worried about her future fate, her first attempts to carve out a career deliberately frustrated by the very family members who should have given encouragement — her mother and her eldest brother Jacob. By chance, her allies — her father and William — could be of only limited support during her formative years; William was in England, while her father was absent with the regiment for long periods, then absorbed in the struggle against ill-health that ended with his early death. She grew up careworn, old beyond her years.

William it was who rescued her, and to repay him by running his household and promoting his current ambitions — in music, or whatever — became her purpose in life. At first this seemed compatible with making a career for herself, as singer or teacher or even harpsichordist, and so achieving financial independence at last. But not for long. The defining moment came when she was approached after she was the leading soprano in a performance of *Messiah*, and invited to sing in Birmingham, on her own account and not as an adjunct to William. In declining the invitation she defined her priorities: William first, career second. Yet, being human and not altogether logical in her reasoning, it rankled that as their household was taken over by William's obsession with manufacturing telescopes, her time for practice diminished and her singing deteriorated. Whom to blame? Not William, of course. It was, she decided, the precious hours she lost in the constant struggle to ensure that the servants she employed were both honest and hardworking. Writing of the "huge blier-eyed Woman" she had had to dismiss reminded her of how much time she had spent supervising staff, time when she could have been practising her singing:

> And in this place I will remark that very seldom I have been so fortunate as to meet with gratitude or good will in any [servants] I have had to deal with, and many times I could not help thinking but that it was owing to a natural antipathy the lower class of the English have against foreigners. In short I have been throughout annoyed and hindered in my endeavours at perfecting myself in any branch of knowledge by which I could hope to gain a creditable livelihood, on account of continual interruption in my practice by being obliged to keep order in a family on which I was myself a dependant.[9]

In her mind it was the servants, not William's astronomical work, that had been responsible for the decline in Caroline's musical career, which took place long before the discovery of Uranus brought it to a definitive end. The move from Bath had only confirmed what they both had known, that circumstances had already conspired against Caroline the soprano.

But her primary vocation — to serve William — was intact. Her golden years were the 1770s and '80s, music and astronomy in Bath followed by astronomy in the environs of Windsor. She had worked all hours, but the work was productive and rewarding — a successful performance of *Messiah*, the netting of a dozen nebulae. William had rescued her from the scullery in Hanover, William was full of plans and ambitions, and William could realize his dreams only with her help. As mistress of his household, and his assistant in astronomy, Caroline was fulfilled in life.

And then William married. She gave up her seat of honour at his table and her primacy in his affections. He now had wealth, a loving wife, a gifted son, a comfortable home, long holidays. Soon she became almost nomadic, forever moving from one set of rooms to the next. It was small compensation that at last she was salaried with money to spend as she wished. Indeed, how to justify her salary itself became a worry as William's hours at the telescope were reduced almost to vanishing.

Yet William's married life began with Caroline comfortably installed in her own cottage next to his house. She destroyed her diary for this difficult period; but in her voluminous and affectionate later correspondence with "Lady Herschel" there is no slightest hint that there had ever been overt trouble between them. Yet for some reason she became "detached from the family circle", and moved out, to begin a wretched and lonely existence in temporary and unsatisfactory lodgings. Small wonder that she was by nature a person whose bottles were half empty rather than half full.

The commitment to serve William that gave meaning to her life survived her demotion and self-inflicted exile from his household. Indeed, towards the end, the fostering of his reputation became an all-consuming passion. Her remarks deprecating her own contributions have to be read in the light of this: "I did nothing for my brother but what a well-trained puppy dog would have done, that is to say, I did what he commanded me."[10] It is a partial truth, the other side of South's coin; but a truth nonetheless. The verdict of Agnes Clerke, *doyenne* of woman historians of astronomy, is as valid today as when she wrote a century ago. First the negative:

> Caroline Herschel was not a woman of genius. Her mind was sound and vigorous, rather than brilliant. No abstract enthusiasm inspired her; no line of inquiry attracted her; she seems to have remained ignorant even of the subsequent history of her own comets. She prized them as trophies....[11]

Then the positive:

> Her persistence was indomitable, her zeal was tempered by good sense; her

endurance, courage, docility, and self-forgetfulness went to the limits of what is possible to human nature. With her readiness of hand and eye, her precision, her rapidity, her prompt obedience to a word or glance, she realised the ideal of what an assistant should be.[12]

John would have agreed:

> She had nothing of mathematical genius, but her extraordinary powers of application in long-continued effort, with her extreme accuracy in all she did, were of as great practical value to my Father as if she had had far greater mathematical knowledge.[13]

On another occasion he wrote:

> She was attached during 50 years as a second self to her Brother, merging her whole existence and individuality in her desire to aid him to the entire extent and absolute devotion of her whole time and powers.[14]

Her role in the great partnership had been exactly what was required. William had enough curiosity, inventiveness and originality for the two of them. He needed a dedicated, reliable and indefatigable assistant, and it was such an assistant that he collected from the family home in Hanover in 1772.

In a letter to John commenting on Baron Fourier's Paris *éloge* in praise of William, Caroline for once allowed herself a hint of credit:

> Of Alex and me can only be said that we were but tools and did as well as we could; but your Father was obliged to turn us first into those tools with which we could work for him; but if too much is said in one place [in praise of me], let it pass; I have perhaps deserved it in another by perseverance and exertions beyond female strength! well done![15]

Well done indeed.

A Note on Sources

The primary source for students of William, Caroline and John Herschel is the treasury of manuscripts held by the Royal Astronomical Society and listed by J. A. Bennett in *Memoirs of the Royal Astronomical Society*, lxxxv (1978). As a service to scholars the papers have been microfilmed in 24 reels and are available for purchase; copies are already to be found in major institutions in various parts of the world.

The Royal Society library holds a large collection of letters of John Herschel. These and other letters of John, some of them to and from Caroline, have been listed and summarized by Michael J. Crowe and colleagues in *A Calendar of the Correspondence of Sir John Herschel* (Cambridge, 1998).

A sale of Herschel manuscripts and other items took place at Sotheby's in London on 4 March 1958 and the manuscripts are now scattered world-wide. The most important collection is in the Harry Ransom Library of the Humanities Research Center of the University of Austin at Texas. It is not, however, widely known that many of the manuscripts now in Austin and elsewhere were microfilmed before export and these microfilms are available for consultation in the British Library. Of these, William's papers to the Bath Philosophical Society and the brief autobiographical memoir he sent to Charles Hutton in 1784 (Lots 475–7, microfilms M/539–41) have been published. Lots 478 (drafts of essays by William), 482 (papers relating to William's life, including a list of telescopes sold), 484 (astronomical tables in Caroline's hand), 485 (Caroline's commonplace book with questions discussed over breakfast), together with the very important autobiographical memoirs and diaries by Caroline sold as Lot 483, are included in the seven reels making up M/588. Also in M/588 are a variety of papers relating to John (Lots 489, 495–7, 499, 501, 503, 504, 514, 517 and 527). Finally, John's drafts of lectures (Lot 511) are in M/542, his notes on acoustics (Lot 507) in M/552, and his notes on chemistry (Lot 513) in M/553. Further details of these microfilms are to be published in *Journal for the History of Astronomy* in 2003.

All these papers were of course available to Mrs John Herschel, author of *Memoir and Correspondence of Caroline Herschel* (London, 1876, 1879; reprinted in 2000 by the William Herschel Society, 19 New King Street, Bath, BA1 2BL); to J. L. E. Dreyer for his important introduction to *The Scientific Papers of Sir William Herschel* (2 vols,

The house and cottage at Slough, from a camera lucida drawing by John Herschel (illustration intended by Lady Lubbock for *The Herschel Chronicle* and in the typescript as submitted to the press, but omitted from the volume as published).

London, 1912); and to Constance A. Lubbock, author of *The Herschel Chronicle* (Cambridge, 1933; reprinted in 1997 by the William Herschel Society). The typescript of *Chronicle* as submitted to Cambridge University Press (and so different from the published version as to be in places almost unrecognizable) is held by the Society and is a valuable resource.

William spent no more time with pen in hand than was necessary (though his "Memorandums from which an historical account of my life may be drawn", RAS MS W.7/8, is important and reliable), and it was Caroline who was the chronicler of the family. She wrote two accounts of her early life: one in the 1820s (cited as CH1) and the other in the 1840s (CH2); the originals are now in Austin and editions of both are in preparation. These autobiographies, together with her memoranda and her later correspondence with John as published in Mrs John Herschel's *Memoir*, are the starting point for any biographical study of William and Caroline.

Notes and References

Abbreviations

Bennett   "'On the Power of Penetrating into Space': The Telescopes of William Herschel", *Journal for the History of Astronomy*, vii (1976), 75–108.

CH1   Caroline's autobiography written in the 1820s.

CH2   Caroline's autobiography written in the 1840s.

Chr   *The Herschel Chronicle: The Life-story of William Herschel and his Sister Caroline Herschel*, ed. by Constance A. Lubbock (Cambridge, 1933).

Clerke   *The Herschels and Modern Astronomy*, by Agnes M. Clerke (London, 1895).

Dreyer   *The Scientific Papers of Sir William Herschel*, ed. by J. L. E. Dreyer (2 vols, London, 1912).

George   *The Later Correspondence of George III*, ed. by A. Aspinall, i (Cambridge, 1962).

Holden   *Sir William Herschel: His Life and Works*, by Edward S. Holden (London, 1881).

James   "Concert Life in 18th Century Bath", by Kenneth E. James (thesis, London University, 1987).

JHA   *Journal for the History of Astronomy*

Maurer   "A Compendium of All Known William Herschel Telescopes", by Andreas Maurer, *Journal of the Antique Telescope Society*, no. 14 (1998), 4–15.

Mem   *Memoir and Correspondence of Caroline Herschel*, by Mrs John Herschel, 2nd edn (London, 1879).

PT   *Philosophical Transactions of the Royal Society*

Slough   *The History of Slough*, by Maxwell Fraser (Slough, 1973).

Turner   *Science and Music in 18th Century Bath*, by A. J. Turner (Bath, 1977).

WH   William Herschel

Citations of the form XVI, 19 (= chapter XVI, page 19) are to the original typescript of *Chr*.

Introduction

1 E. F. MacPike, *Hevelius, Flamsteed and Halley* (London, 1937), 4–16.

2 *Dictionary of Scientific Biography*, vii, 374.

3 CH1, 58.

Hanoverian Prologue

1 CH1, 23.

2 CH2, 13.

3 I, 22.

4 *Mem*, 10–11.

5 CH2, 13.

Chapter 1

1 *Mem*, 1 gives August. However, Jürgen Hamel tracked down a photocopy of the original entry in Stadtarchiv Hannover, see his "Ein Beitrag zur Familiengeschichte von Friedrich Wilhelm Herschel", Gauss-Gesellschaft E. V. Göttingen, *Mitteilungen* 26 (1989), 99–103. In this entry, Anna's second name is spelt "Ilsa".

2 On Isaac's ancestry and early life, see his Autobiography as reproduced in *Chr*, 2–3; for the earlier Herschels, see *Chr*, 1.

See also CH1, i, 1–4, *Mem*, 1–2, and RAS MS W.7/8.

3 *Mem*, 11.

4 *Chr*, 10.

5 CH2, draft, i, 12.

6 *Mem*, 20.

7 *Chr*, 2.

8 *Chr*, 2. The brothers, however, did not keep in touch in later life, CH1, i, 1.

9 As we shall see, all five became members of the

Court orchestra at Windsor.

10 *Chr*, 2.

11 *Ibid*.

12 *Ibid*.

13 CH1, i, 3.

14 *Chr*, 3.

15 *Chr*, 3.

16 Isaac arrived in Hanover on 4 August and obtained a post in the band on the 7th, CH2, i, 1.

17 They were married on 12 October 1732, as proved by the official register of the Garrison Church (see above), and Sophia was born on 12 April of the following year. That Anna was pregnant with Sophia was a secret kept from the children. *Mem*, 1, would have us believe that the marriage took place in August, in which case Sophia might well have been a premature baby conceived in wedlock; even so, while the birthdays of the other five surviving children are listed there, that of Sophia is curiously omitted. By contrast, *Chr*, citing Isaac, gives Sophia's birthday but only the year of the marriage.

18 "I soon arrived to such a degree of perfection, especially in arithmetic, that the Master of the school made use of me to hear younger boys say their lessons and to examine their arithmetical calculations", RAS MS W.7/8, 6.

19 RAS MS W.7/8, 6.

20 CH1, i, 13.

21 "I never could remember the multiplication table, but was obliged to carry always a copy of it about me", Caroline to John Herschel, 1 September 1840, *Mem*, 316.

22 CH1, i, 9.

23 CH1, i, 9.

24 CH1, i, 9. "In this place it must be remembered how enthusiastically fond my Father was of Music, and his desire in consequence was, to see his children arrive at that eminence in this his favourite science, which he himself had not had the opportunity or time to attain."

25 Even Johann Heinrich, who died young, had shown talent by the time he was five, CH1, i, 10.

26 "... when our dear Father had leisure to look over, and correct our lessons when coming from school", CH2, draft, i, 12; "... my brother W^m and his Father often were arguing with such warmth that my Mother's interference became necessary when the names Leibnitz, Newton and Euler sounded rather too loud of the repose of her little ones", CH1, i, 17.

27 I borrow this phrase from J. B. Sidgwick.

28 CH2, i, 2.

29 CH2, i, 2–3. Isaac says he was in Hanover on leave from 10 January to 4 April 1744.

30 *Chr*, 6.

31 CH1, i, 11.

32 CH2, i, 16, note.

33 William wrote to Jacob on 16 November 1761: "I wish with all my heart that my father could get the situation now vacant which he well deserves, having served the King so long and grown old in his service", II, 23. This is surely a reference to the Court orchestra.

34 One at least of her brothers was a farmer, CH1, i, 8. Anna had in all three brothers and two sisters, and a great many nephews and nieces, CH1, i, 4.

35 CH1, i, 7.

36 CH1, i, 4. He joined the band on 1 May 1753, RAS MS W.7/8, 7.

37 *Chr*, 5.

38 CH1, i, 10.

39 RAS MS W.7/8, 7.

40 RAS MS C.4/1.2.

41 *Ibid*.

42 CH2, i, 7.

43 CH2, i, 8.

44 *Chr*, 6, 7; I, 13. Griesbach's Christian names, curiously absent from Herschel family papers, are given in the entries for his sons in vol. vi of *A Biographical Dictionary of Actors, Actresses, Musicians, Dancers, Managers and Other Stage Personnel in London, 1660–1800*, by Philip. H. Highfield, Jr, Kalman A. Burnim and Edward A. Langhams (Carbondale and Edwardsville, 1978). Information in these entries is not available elsewhere, but must be used with caution.

45 RAS MS C.4/1.2.

46 CH1, i, 15.

47 *Ibid.*

48 *Ibid.*

49 CH2, i, 7.

50 RAS MS C.4/1.2.

51 *Chr*, 34.

52 Her adult height has been quoted as 4ft 3ins, *Chr*, 170, but this is an error. Her surviving dress is just under 47 inches from the back of the neck to the hem, and from the back of the neck to the top of the head is typically about 10 inches. The distance from the hem to the top of her head, therefore, was about 57 inches, to which must be added whatever distance there was between the hem and the sole of her feet. Even if the hem touched the floor, her height was 4ft 9inches.

53 CH1, i, 14; *Chr*, 3.

54 CH1, i, 64.

55 *Mem*, 20.

56 CH2, i, 15.

57 *Mem*, 20.

58 CH2, i, 23.

59 CH2, i, 24.

60 CH2, i, 26: "[Jacob] had given his consent on condition of my not thinking on going into service or taking in work."

61 VI, 3.

62 *Mem*, 20.

63 *Mem*, 20.

64 Caroline to John's wife, 24 September 1838, *Mem*, 300.

65 CH1, i, 14. Since the figures cited are consistent, I follow the chronology of CH2, i, 2, according to which "Frans" was born 5 February 1752, lived 2 years and 7 weeks, and died 26 March 1754. According to CH1, i, 4, however, "Frantz" was born on 13 May 1752 and lived 1 year, 11 months and 23 days, in which case he died on 6 May 1754.

66 CH2, i, 8.

67 CH2, i, 9.

68 William says it was "Soon after the great Earthquake" of 1 November, RAS MS W.7/8, 7.

69 CH1, i, 18–19.

70 According to CH1, i, 20. CH2, i, 9 limits the arrangement to Isaac himself.

71 CH2, i, 9.

72 *Ibid.*

73 CH2, draft, i, 11.

74 CH2, i, 9.

75 J. W. Fortescue, *A history of the British Army*, ii (London, 1910), 296.

76 RAS MS W.7/8, 7.

77 CH1, i, 22.

78 *Chr*, chap. 2.

79 CH1, i, 19; CH2, i, 8.

80 CH2, draft, i, 9.

81 RAS MS W.7/8, 8.

82 The return of the Hanoverian troops was announced in Parliament on 2 December, Fortescue, *op. cit.*, 305.

83 CH1, i, 21.

84 I, 6.

85 CH2, draft, i, 11.

86 CH2, draft, i, 12.

87 RAS MS W.7/8, 8.

88 J. B. Sidgwick, William Herschel (London, 1953), 21.

89 RAS MS W.7/8, 9.

90 CH2, draft, i, 13 says William's return "must have been about the end of August", but it was within a very few days of the battle, and certainly before the French entered Hanover on 10 August.

91 RAS MS W.7/8, 9.

92 R. A. Savory, *His Britannic Majesty's Army in Germany during the Seven Years War* (Oxford, 1966), 45.

93 RAS MS W.7/8, 9.

94 Savory, *op. cit.*, 39.

95 CH2, draft, i, 14.

96 But not until 1762. The document is cited in full by Dreyer, i, p. xvi.

97 CH2, draft, i, 14.

98 RAS MS W.7/8, 9.

99 CH2, draft, i, 14: "he was put in Arest by way of enforcing the return of the Deserter."

100 CH1, i, 24.

101 CH2, draft, i, 14.

102 RAS MS W.7/8, 9.

103 CH2, i, 14. CH1, i, 24: "My brother keeping himself so carefully from all notice was undoubtedly to avoid the danger of being pressed, for all unengaged young men were forced in the service. Even the Clergy without they had livings were not exempt."

104 CH1, i, 24. George Ludolph Jacob Griesbach was born on 10 October and baptised three days later, *A Biographical Dictionary*..., vi, 358.

105 CH2, draft, i, 15.

106 Caroline to John, April 1827, *Mem*, 211–12.

107 Savory, *op. cit.*, 42.

108 Savory, *op. cit.*, 62. The approximate date of the French abandonment of Hanover was kindly confirmed for me by Arndt Latusseck through a study of the Hanoverian state archives.

109 *Chr*, 11.

110 II, 23.

111 CH1, i, 42.

112 *Ibid.*

113 IV, 2.

114 CH2, draft, i, 15.

115 CH2, draft, i, 16.

116 CH1, i, 20.

117 CH1, i, 15.

118 CH2, draft, i, 16.

119 IV, 3.

120 I, 15.

121 CH2, draft, i, 17.

122 CH1, i, 35. Sophia's baby was Charles, the least-known of her five sons.

123 CH2, draft, i, 17.

124 CH2, i, 20.

125 CH1, i, 33.

126 CH2, draft, i, 19.

127 *Chr*, 11; CH1, draft, i, 20.

128 CH1, i, 38. Dreyer, i, p. xvi, cites his discharge as 11 May 1760; according to Caroline, CH2, draft, i, 21, he accompanied the Guards when they left Paterborn on the 12th and reached Coppenbrügge on the 14th, where he had to rest before being strong enough to continue to Hanover. *Cf.* CH1, i, 26.

129 III, 24: "I have for some time been thinking of leaving off professing music, and the first opportunity that offers, I shall really do so."

130 William on one occasion played the continuo rather than be second fiddle, III, 27.

131 II, 2.

132 RAS MS W.7/8, 10.

133 *Ibid.*

134 II, 2.

135 IV, 2.

136 I, 5.

137 CH1, i, 60.

138 CH1, i, 34.

139 *Ibid.*

140 CH2, i, 17.

141 CH1, i, 37.

142 CH1, i, 38. By 1764 "my eldest brother was so constant in his attendances on one of the ladies I have mentioned before (Fraulein Westernholtz) that he was seldom at home except to dress and take his meals", CH1, i, 48.

143 CH2, i, 19.

144 *Ibid.*

145 CH2, i, 24.

146 CH1, i, 43.

147 CH1, i, 48.

148 CH1, i, 50.

149 CH1, i, 51.

150 CH1, i, 54.

151 *Ibid.*

152 CH2, i, 20.

153 CH1, i, 64.

154 CH1, i, 46. Isaac did not have the English to read the long letters William was sending Jacob.

155 He arrived on 2 April and left on the 15th, *Mem*, 16–18.

156 CH2, draft, i, 25.

157 CH2, i, 22.

158 CH1, i, 60.

159 CH1, i, 55.

160 CH1, i, 58.

161 CH1, i, 61.

162 Jacob played the oboe in the Guards Band and became organist of the Garrison Church in Hanover, but he was primarily a violinist. William played the oboe, clarinet, violin and organ, and sang as a tenor soloist. Alexander was trained in Hanover in the violin and oboe; in Bath he early on distinguished himself with the clarinet, yet it was as a cellist that he achieved his greatest fame. On all these, see James, 690–711. Dietrich, whose hobby was natural history, was a violinist, and he is spoken of by William as "an eminent Musician, and well known as a scientific member of several

Academies", RAS MS W.7/8, 5.

163 CH1, i, 62.
164 CH1, i, 64.
165 Caroline to John, 25 September 1827, *Mem*, 218.
166 CH1, i, 65.
167 CH2, i, 28.
168 CH1, i, 65.
169 *Ibid.*
170 CH2, i, 28.
171 CH1, i, 65.
172 CH1, i, 66.
173 CH1, i, 65.
174 CH2, i, 27. Caroline cites the place as Görde. I am grateful to Arndt Latusseck for the identification.
175 "... on account of the enormous expense of living as boarders in the Foresters Family", CH2, i, 27.
176 CH1, i, 67.
177 CH1, i, 66.
178 CH2, i, 28.
179 CH2, i, 27.
180 CH1, i, 67.
181 Caroline to John's wife, 24 September 1838, *Mem*, 299.

Chapter 2

1 *Chr*, 48.
2 *Ibid.*
3 This account is based on VI, 13 et seq.
4 CH2, i, 29–30.
5 *Chr*, 49.
6 *Chr*, 38.
7 J. Haslewood, *The Secret History of the Green Room, containing authentic and entertaining memoirs of the actors and actresses of the three Theatres Royal*, 3rd edn (London, 1793), ii, Appendix, p. xv. This little-known episode in William's life was elucidated by James, 469–71. Elizabeth Harper may be the "Miss Hooper" who sang in William's very first concert in Bath on New Year's Day, 1767, according to S. Derrick, *Letters written by Samuel Derrick* (London, 1767), ii, 102.
8 *Chr*, 43.
9 *Chr*, 49.
10 *Chr*, 52.
11 CH2, ii, 1–2.
12 *Ibid.*
13 *Ibid.*
14 CH1, ii, 1.
15 *Chr*, 50.
16 CH1, ii, 2.
17 VI, 16.
18 CH2, ii, 11.
19 CH2, ii, 4.
20 *Chr*, 49.
21 CH2, ii, 4.
22 CH2, ii, 5.
23 *Chr*, 50.
24 CH1, ii, 6.
25 CH2, ii, 5.
26 CH1, ii, 2.
27 CH2, ii, 7.
28 CH1, i, 7; CH1, ii, 3; CH2, ii, 5. Johann Wilhelm Griesbach was born on 10 January (*A Biographical Dictionary...*, vi, 360), so his father's death took place after that date.
29 Caroline to Mary Herschel, 14 October 1824, RAS MS C.4/1.2.
30 CH1, ii, 3.
31 *Chr*, 51–52.
32 Turner, 31.
33 On the 29th, RAS MS W.7/8, 26.
34 On the 4th, RAS MS W.7/8, 27.
35 James, 702.
36 RAS MS W.7/8, 24–26; James, 701.
37 James, 218, citing Fanny Burney.
38 James, 416.
39 James, 703–4.
40 Ian Woodfield, *The Celebrated Quarrel Between Thomas Linley (Senior) and William Herschel: An Episode in the Musical Life of 18th Century Bath* (Bath, 1977).
41 RAS MS W.7/8, 31.
42 CH1, ii, 2. Caroline may mean by this that she was accompanying her own singing on the harpsichord.
43 CH2, ii, 6.

44 CH2, ii, 7–8.

45 CH2, ii, 8.

46 CH1, ii, 3. In CH2, ii, 6, it becomes "a Bason of Milk or Tea", and William's exhaustion is in the aftermath of the season rather than during it.

47 CH1, ii, 3.

48 CH2, ii, 6–7.

49 *Chr*, 5.

50 CH1, i, 47, 1 April 1764, the eve of William's arrival home.

51 *Chr*, 50.

52 *Chr*, 43; V, 23–24. In 1769 he earned £316, in 1770 £352, and in 1771 "nearly 400 pounds".

53 *Chr*, 60.

54 It was first published in London in 1756 and went through seventeen editions.

55 On Ferguson see J. R. Millburn, *Wheelwright of the Heavens* (London, 1988), and the brief entry by Laurens Laudan in *Dictionary of Scientific Biography*, iv, 565–6.

56 *Chr*, 60.

57 On this see the author's *William Herschel and the Construction of the Heavens* (London, 1963), 22–23.

58 *Mem*, 35.

59 Robert Smith, *Harmonics, or the Philosophy of Musical Sounds* (Cambridge, 1749). William's copy was sold as Lot 446 of the auction at Sotheby's in March 1958. He had intended to write a Treatise on Harmony as early as 1758 "and began to collect materials for that purpose", RAS MS W.7/11, 20.

60 Robert Smith, *A Compleat System of Opticks* (Cambridge, 1738).

61 CH1, ii, 3.

62 RAS MS W.7/8, 29.

63 CH1, ii, 3–4.

64 RAS MS W.7/8, 30.

65 CH2, ii, 8. Cf. CH1, ii, 4.

66 *Chr*, 65.

67 VII, 13.

68 CH2, ii, 9.

69 Footnote added in the rough draft of CH2 (in private possession), but omitted in the fair copy. She seems to be saying that "the compliments ... were at last carried of[f] by the wicked pilfering wretches",

but if her metaphor is obscure, her disapproval is clear.

70 *Chr*, 43.

71 *Chr*, 53.

72 *Chr*, 55.

73 In April according to CH2, ii, 17; at midsummer according to *Mem*, 37.

74 This is the address used by William on a letter to Caroline from Hanover dated 22 August 1777, in private possession. It has often been assumed that the house was at the Walcot turnpike, and indeed Caroline speaks of it as "situated near Walcot Turnpike, which at that time was without the district of the Mayre and free from Buildings till near Bath-Easton" (CH1, ii, 8). But the turnpike would be surprisingly far out of town: New King Street is 7 minutes walk from the Abbey, Walcot Parade 13, and the turnpike no less than 23 (Richard Phillips, private communication, 2002). William simply refers to "House at Walcot" in his autobiographical memoranda (RAS MS W.7/8, 31).

75 CH2, ii, 14.

76 CH2, ii, 21. The house was close enough to town for Alexander to come "dayly to spend his leisure hours with us, or in the workshop".

77 CH2, ii, 16.

78 CH1, ii, 8.

79 CH2, ii, 14.

80 *Ibid.*

81 On William's telescopes in general, see Bennett.

82 CH1, ii, 8–9.

83 CH2, ii, 15.

84 CH2, ii, 7.

85 CH2, ii, 15.

86 *Ibid.*

87 *Ibid.*

88 *Ibid.*

89 RAS MS W.7/13.

90 *Ibid.*

91 *Mem*, 324–5.

92 The book, which has a hundred or so pages, many of them blank, was auctioned by Sotheby's on 3/4 March 1958 as Lot 485, and it is now at the University of Texas at Austin. Dr Bradley Schaefer examined

it, and kindly supplied the author with a detailed summary. In the article on Caroline for the *Dictionary of National Biography*, Agnes Clerke well describes her commonplace book as containing "a miscellaneous jumble of elementary formulae, solutions of problems in trigonometry, rules for the use of tables of logarithms, for converting sidereal into solar time, and the like". Spherical triangles occur on pages 55 and 64.

93 RAS MS W.5/12/1, 44. Reflectors needed duplicate primary mirrors because the metal tarnished while in use under the damp night sky. By having duplicate mirrors, the observer could use one in the telescope while the other was being repolished.

94 RAS MS W.5/12.1, 45: "I prepared to repolish my brother John's 3-feet speculum." Alexander's first name was Johann.

95 John Marsh, manuscript memoirs, ix, 753–4 [Cambridge University Library, Add MS 7757], tells of sitting next to William at supper in May 1782. "... entering into some discourse with him upon astronomy, which he then applied to much more than music, he told me of his being then at work upon a mirror for a large telescope, of the magnifying powers of which he was very sanguine in his hopes, and which proved to be the one which soon afterwards brought him to the notice of his majesty, and occasioned his removal to Windsor. His sister and his brother, who played the principal violoncello at Bath, was as fond of astronomy as himself and all used to sit up, star-gazing, in the coldest frosty nights." Marsh is totally confused about William's telescopes, and is doubtless equally confused about the current roles of Caroline and Alexander. I am indebted to Kenneth James for this reference.

96 *Chr*, 68, 72.

97 Turner, 33 and ref. 57.

98 Turner, 38.

99 James, 214.

100 As the poster shows, for the *Messiah* performed at "Mr. Herschel's Benefit-Concert" on 15 April 1778 all the seats cost five shillings. Tickets could be bought at William's home, as well as elsewhere. For the performance of *Alexander's Feast* on 15 March 1780, tickets cost between two and five shillings (Turner, 64).

101 Around 1810 William's gardener, Cock, received twelve shillings a week, Accounts (in private possession), entry for 24 March 1810.

102 CH2, ii, 16.

103 CH1, ii, 10.

104 CH2, ii, 17.

105 CH2, ii, 16. He did however find time to set up a mural telescope: "... there could only remain a small portion of time for spending in his workshop besides of contriving to fix a wall-telescope for the transit of a star (I think it was Capella) and inspecting the making of a time-piece by Field a famous watchmaker in Bath." This incursion by William into traditional astronomy has received little attention.

106 CH2, ii, 17.

107 James, 695.

108 CH1, ii, 9.

109 *Chr*, 56.

110 Frank Brown, *Caroline Herschel as a Musician* (Bath, 2000), 9. Caroline was second principal to Elizabeth Mahon.

111 RAS MS W.7/8, 32.

112 *Chr*, 56.

113 CH2, ii, 18.

114 *Ibid.*

115 CH1, ii, 10.

116 CH2, ii, 19.

117 William learned of Dietrich's bid for freedom on 30 July 1777 and left for London immediately; he arrived in Holland on 3 August, and reached Hanover on the 11th (VII, 16–18).

118 VII, 18.

119 *Chr*, 70–71.

120 *Chr*, 71.

121 CH1, ii, 11. There is nowhere any hint that either William or Alexander or Caroline ever reproved Dietrich for the immense trouble he had caused, which suggests that they understood all too well what had led him to attempt this break for freedom.

122 VII, 22.

123 Heinrich emigrated in his early twenties and died soon after at Charleston, VII, 23.

124 James, 706.

125 RAS MS W.7/8, 32. The last concert was on 10 April 1778.

126 "We had weekly concerts at Bristol; at the lower Rooms at Bath, and also at the New Rooms", RAS MS W.7/8, 32.

127 James, 218.

128 CH1, ii, 12.

129 James, 696.

130 RAS MS W.7/8, 33, memo for January 1779.

131 RAS MS W.7/8, 33.

132 Thomas Hornsby to William, 22 December 1774, RAS MS W.1/13.H.23, speaks of "Since I had the pleasure of seeing you...". Hornsby goes to great pains to explain the geometry and physics of the immersion and emersion of the satellites of Jupiter, and he discusses the disappearance of Saturn's ring. Their meeting must have been brief, for Hornsby writes: "As you seem very fond of the science of Astronomy, may I presume to ask what instruments you have, and particularly how you gain your Time [of observations]." In a rare error, *Chr*, 75 puts their meeting at about 1777.

133 *Chr*, 75. "Dr Lysons" was probably Samuel Lysons (1763–1819), antiquary and later Vice-President of the Royal Society.

134 Edward Pigott, *First Astronomical Journal* (Pigott MSS, North Yorkshire County Council), entries for 12 April 1778 *et seq.* *Cf.* Pigott to William, 24 August 1817, RAS MS W.1/13.P.42, where he reminds William that it is 39 years since they observed together at New King Street.

135 They moved on 29 September 1777, RAS MS W.7/8, 32. The house in New King Street is now a shrine for historians of astronomy and the home of the William Herschel Society.

136 Personal communication, 2002, from Michael Tabb of Bath, based on his detailed analysis of surviving rate books in Bath City Records Office. The rate collectors went first along one side of the street and then back along the other, so that their entries are not in the numerical

order of the houses (and this has misled earlier investigators into thinking that William lived at no. 5).

137 CH1, ii, 14.

138 Turner, 82, cites the Secretary's Journal as giving 27 December 1779 for the meeting between Thomas Curtis and Edmund Rack (the future Secretary) at which the two men agreed to try and promote a society.

139 *Chr*, 73.

140 CH1, ii, 14–15.

141 *Chr*, 73.

142 Turner, 82.

143 This is the conclusion of Trevor Fawcett of Bath (personal communication, 2002). Watson, who lived in no. 23 of the famous Royal Crescent, had recently bought a plot from Charles Hamilton, a landscape gardener living at no. 14.

144 Most of these are published in full in Dreyer's edition of William's collected papers.

145 *Chr*, 77.

146 WH, "Astronomical Observations Relating to the Mountains on the Moon", PT, lxx (1780), 507–26.

147 RAS MS W.7/8, 33.

148 RAS MS W.7/8, 34; James, 706.

149 RAS MS W.7/8, 34.

150 Reproduced in Turner, 64.

151 *Chr*, 57.

152 Dreyer, i, p. xxix.

153 WH, "Catalogue of Double Stars", PT, lxxii (1782), 112–62, IV.1.

154 Sir Joseph Banks to William, 15 March 1782, RAS MS W.1/13.B.4.

155 CH1, ii, 14.

156 This very familiar story has been discussed many times. See for example Dreyer, i, pp. xxviii–xxxi, and J. A. Bennett, "Herschel's Scientific Apprenticeship and the Discovery of Uranus", in G. E. Hunt (ed.), *Uranus and the Outer Planets* (Cambridge, 1982), 35–53.

157 On events following the discovery see for example Simon Schaffer, "Uranus and the Establishment of Herschel's Astronomy", *JHA*, xii (1981), 11–26.

158 *Chr*, 79.

159 *Gazette Littéraire*, June 1781, cited Holden, 55, from *Berliner Jahrbuch*, 1784, 211.

160 Ch. Messier to William, 29 April 1781, RAS MS W.1/13.M.97.

161 *Journal Encyclopédique*, cited Holden, 55.

162 *Connaissance des Temps* for 1784, according to Holden, 55.

163 RAS MS W.7/8, 34. William merely remarks on the dates of the oratorios, but the Wednesday/Bath, Friday/Bristol pattern was well established.

164 Easter Sunday fell on 15 April.

165 RAS MS W.7/8, 34.

166 CH1, ii, 15.

167 CH1, ii, 15–16.

168 CH1, ii, 17.

169 *Ibid.*

170 *Chr*, 89. William did not regret this as a small crack had appeared in the bottom of the mould during casting and as a result there was "a great deficiency in one side of the mirror".

171 CH1, ii, 18.

172 *Chr*, 95.

173 CH1, ii, 17.

174 *Chr*, 96. His nomination speaks of his "having communicated several papers to the Royal Society, particularly those relating to the present comet [Uranus], which he first observed" (Royal Society archives).

175 Watson, in a letter of 25 December 1781 (RAS MS W.1/13.W.13), hints at William's "pecuniary interest".

176 S. C. T. Demainbray to William 12 August 1781, RAS MS W.1/13.D.14.

177 George ordered Jacob's annual salary to be increased by 100 thalers, IV, 3.

178 VI, App 4.

179 *Chr*, 112.

180 *Ibid.*

181 *Ibid.*

182 *Chr*, 113.

183 RAS MS W.7/8, 34. William dates this remark to April, but, as we shall see, he already knew in March that he was to be summoned.

184 RAS MS W.7/8, 34.

185 CH1, ii, 17.

186 James, 236.

187 CH1, ii, 19.

188 Cited by James, 236.

189 James, 237.

190 Brown, *Caroline Herschel as a Musician*, 10.

191 CH1, ii, 19.

192 On the Monday according to William (RAS MS W.7/8, 35), the Tuesday according to Caroline, CH, ii, 20.

193 CH1, ii, 20.

194 *Chr*, 115.

195 John Walsh to William, 10 May 1782, RAS MS W.1/13.W.5; *Chr*, 113.

196 *Chr*, 114.

197 James, 238.

198 X, 5.

199 *Chr*, 115.

200 *Chr*, 115–16.

201 Caroline to Lady Herschel (wife of Caroline's niece John), 3 February 1842; *Mem*, 320–2.

202 "A 3½ feet and a 10 or 12 feet Achromatic by Dollonds and a Short's reflector were compared with my instrument", RAS MS C.3/1.1.

203 X, 11–13, quotation on p. 11.

204 *Chr*, 118.

205 "He [the King] has done on his side everything to show his partiality towards you and it cannot be expected that he should condescend to offer, before he knows the offer will be accepted", William Watson, Jr, to William, 29 June 1782, RAS MS W.1/13.W.18.

206 The King uses the word 'salary' in approving the payments for the year 1800, papers in private possession.

207 RAS MS W.7/8, 35.

208 "A Letter from William Herschel to Sir Joseph Banks", *PT*, lxxiii (1783), 1–3.

209 *Chr*, 133.

210 *Chr*, 115.

211 *Chr*, 128.

212 RAS MS W.7/8, 35.

213 George, Letter 656, 26 Feb 1791: "At the same time I beg the favor of you to tell me whether I shall send the bill for these five telescopes to you [a royal secretary] to give to the King, or whether you know where it would please his Majesty that I should send it. The King knows very well that astronomers and experimental philosophers are always poor, and, as

the liberal patron of Arts and Sciences, I make no doubt, will not suffer them to be in want."

214 RAS MS W.7/8, 35–36.

215 CH1, ii, 19.

216 CH1, ii, 21–22.

217 CH1, ii, 19. Clerke, 121, quotes her as saying

that her thoughts "were anything but cheerful" on this occasion. If so, she had a premonition of the end of her singing career.

218 CH1, ii, 24. For "impudence" Caroline originally wrote "spirit".

Chapter 3

1 *Chr*, 134. The inn later became the Royal Stag.

2 *Ibid.*

3 From the *Reading Mercury*, advertising the property when William and Caroline left in 1785. The grander house adjoining, "The Lawns", still exists, though much altered. I owe this information to the kindness of Janet Kennish, in whose *Datchet Past* (Chichester, 1999) further information can be found.

4 *Chr*, 135.

5 Nevertheless the business survived as a butcher's shop until the early 1970s, when it was sold (personal comm, 2002, from Janet Kennish).

6 *Chr*, 135.

7 The Visitors Book (currently on loan to the Herschel Museum) shows that the King and the Duke of Clarence were among those who visited Herschel in 1783/84, when he was in tumbledown accommodation at Datchet.

8 RAS MS C.3/1.2, 57.

9 Mrs Beckedorff to William, n.d., RAS MS W.1/13.B.57.

10 For example, 17 April 1796: "The King and two Equerries. His Majesty saw the moon in a 10 feet reflector", memorandum included in Lot 480 of the Sotheby's sale.

11 RAS MS W.4/1.3, f. 217.

12 Published in 1781 as an additional item in *Connaissance des Temps* for 1784.

13 Owen Gingerich, entry on Messier in *Dictionary of Scientific Biography*, ix, 329–31, p. 329.

14 See Michael Hoskin, "William Herschel's Early Investigations of Nebulae: A Reassessment", *JHA*, x (1979), 165–76.

15 RAS MS W.4/1.3, f. 224.

16 *Chr*, 136. The (sidereal) time when a star crosses the meridian yields one of the coordinates of the star.

17 "My brother and sister were with me [when I was first at Datchet], the former on a visit, the latter to be my assistant in astronomy, in which capacity she had already acted at Bath", RAS MS W.7/8, entry for August 1782.

18 Caroline kept "Journals", rough notes of which no. 4 survives (RAS MS C.1/1.4), and "Books of Observations" (C.1/1.1–3), which were fair copies of the Journals. In 1828 she sent John the Books of Observations, plus Journal no. 4 because not everything was transcribed from it. In a Memorandum dated 7 December 1833 (C.4/3, 3r), she says: "in the beginning [!] of the year 1782 I began to sweep the heavens and to note down whatever appeared to be remarkable in quarto books which I called Journals 1, 2, &c and transcribed the same in books called Observations. These books I thought it best to destroy, excepting some fragments which I some 4 or 5 years since sent to my Nephew as waste paper. For, in consequence of my employment at the Clocks and writing Desk, when my Brother was observing, I had no other opportunity for looking out for comets, but when he was absent from home, but this happened so seldom and my sweeps were so broken and unconnected that I could not bear the thought of their rising against me; as besides they contained nothing new but the discovery of 8 comets and a few neb. and cl. of stars." The comets, she says, were communicated to the Astronomer

Royal, and the nebulae and clusters of stars inserted in William's catalogues with her initials.

19 CH1, ii, 28; the transcriptions of this passage in *Mem*, 52, and *Chr*, 149–50, are, as so often, inexact. The sketch of this primitive first sweeper is among the papers of William Henry Smyth in the Museum of the History of Science at Oxford (MS Gunter 36, ff. 124v). It presumably dates from 1843/44 when Smyth was preparing *The Bedford Catalogue*, vol. ii of his *Cycle of Celestial Objects* (London, 1844), and must have been based on information supplied by John Herschel. The sketch of the 1783 'small' sweeper (to distinguish it from the 'large' one William made Caroline in 1791) is also by Smyth (*ibid.*, 125v); it is closely based upon the sketch sent him by John on 21 November 1843 (letter in The National Library of South Africa) and was used by Smyth for the figure on p. 540 of the *Catalogue*. See Margaret Bullard, "My Small Newtonian Sweeper — Where Is It Now?", *Notes and Records of the Royal Society of London*, xvii (1988), 139–48.

20 RAS MS C.1/1.1, pasted in front; XII, 6.

21 RAS MS C.1/1.1, 1.

22 RAS MS C.1/1.1, 1.

23 RAS MS C.1/1.1, 1.

24 RAS MS C.1/1.1, 2.

25 RAS MS C.1/1.1, 2–3.

26 RAS MS C.1/1.1, 4.

27 RAS MS C.1/1.1, 4–5.

28 RAS MS C.1/1.1, 6.

29 *Mem*, 52.

30 RAS MS C.1/1.4.

31 WH, "On the power of penetrating into space by telescopes", *PT*, xc (1800), 49–85, p. 71. Caroline gives some details of the sweeper in her first Book of Observations (RAS MS C.1/1.1, 11). The optics survive to this day in the Historisches Museum in Hanover.

32 *Chr*, 233.

33 Nathaniel Pigott archives of the Royal Astronomical Society, Letter 60.

34 RAS MS C.1/1.1, 16.

35 RAS MS C.1/1.1, 19. The companion is now known as NGC 205. In 1807 Messier published a drawing of the Andromeda Nebula showing its companions M32 and NGC 205, and claiming to have himself seen NGC 205 as long ago as 1773. If so, one wonders why he did not include it in his catalogue. See Kenneth Glyn Jones, *The Search for the Nebulae* (Chalfont St Giles, 1975), 75.

36 RAS MS W.4/1.4, f. 338.

37 On the finished version of the 20-ft, see Bennett.

38 *Chr*, 136–7.

39 Galileo Galilei, *Dialogue of the Great World Systems*, transl. by Giorgio de Santillana (Chicago, 1953), Third Day, 393. The sale of the Herschel Library by Sotheby's in March 1958 included a copy of the Latin edition, *Systema cosmicum* (Leyden, 1699) as Lot 375, and this was presumably the version William used.

40 William rightly credits the method to Galileo in the paper he wrote in 1781, "On the Parallax of the Fixed Stars", *PT*, lxxii (1782), 82–111, to establish the context for his first catalogue of double stars; and by this time he had read Galileo on the subject. He remarks that "This method has also been mentioned by other authors", and discusses its attempted application by Professor Roger Long (1680–1770) of Cambridge, from whose writings he may first have learned of it. On this see my "Stellar Distances: Galileo's Method and its Subsequent History", *Indian Journal of History of Science*, i (1966), 22–29; and "Herschel and Galileo", *Actes du XIe Congrès International d'Histoire des Sciences* (Paris, 1968), iii, 41–44.

41 John Michell, "An Enquiry into the Probable Parallax and Magnitude of the Fixed Stars", *PT*, lvii (1767), 234–64.

42 *Chr*, 87.

43 WH, "Account of the Changes That Have Happened, During the Last Twenty-five years, in the Relative Situation of Double Stars; with an Investigation of the Cause to which They Are Owing", *PT*, xciii (1803), 339–82.

44 Though not quite the first, one of the successful investigators was William's

son John: J. F. W. Herschel, "On the Investigation of the Orbits of Revolving Double Stars", *Memoirs of the Royal Astronomical Society*, v (1833), 171–222.

45 F. W. Bessel, "A Letter of Professor Bessel to Sir John Herschel, Bart.", *Monthly Notices of the Royal Astronomical Society*, iv (1839), 152–61. Although Galileo spoke of double stars, his proposal was in essence to measure the position of a near star relative to a distant star in the same region of the sky, whether or not the two stars were so close as to constitute a double.

46 *Mem*, 53–54.

47 *Chr*, 150–1.

48 WH, "Catalogue of One Thousand New Nebulae and Clusters of Stars", *PT*, lxxvi (1786), 457–99, p. 460.

49 Caroline to John, 1 February 1826, *Mem*, 196.

50 *Mem*, 54.

51 P.-F.-A. Méchain to William, 25 October 1789, RAS MS W.1/13.M.84: "... sa célébrité sera en honneur dans tous les siècles."

52 *Chr*, 137.

53 *Ibid*.

54 Letter of Jean-Hyacinthe de Magellan to J. E. Bode published in Bode's *Jahrbuch*, 1788, 161, and translated in Holden, 78–79.

55 As she explains in CH1, ii, 40, her first list (RAS MS C.2/1.1) did not include stars less than 45° from the Pole, "The apparatus not being then ready for sweeping in the zenith". The final and complete list is C.2/1.2. As William later explained, "By a catalogue in zones the assistant may guide the observer ... who ought to have notice given him of such stars as have their places well settled, in order to deduce from their appearance the situations of other objects that may occur in the course of a sweep. In the year 1783, when I began this kind of observations, no catalogue of stars in zones had ever been published; I therefore gave a pattern to my indefatigable assistant, Carolina Herschel, who brought all the British catalogue into zones of one degree each" (WH, "Description of a Forty-feet Reflecting Telescope", *PT*, lxxxv (1795),

347–409, pp. 395–6).

56 RAS MS C.1/1.1, 31.

57 RAS MS C.1/1.1, 32.

58 CH1, ii, 33.

59 XI, 24.

60 Later Clayhall Farm, Clayhall Lane. The house, on the edge of Windsor Great Park and a mile from the castle, survived until 1979 (E. A. Fanning, "Where Did William Herschel Live?", The William Herschel Society *Bulletin*, no. 58 (1999), 14–16, p. 15).

61 *Chr*, 144. Information from Cdr Fanning.

62 XI, 25.

63 Sweeps nos 549 and 550.

64 RAS MSS W.2/3.5.

65 The house was later renamed "Ivy House", and in the nineteenth century became "Observatory House". It survived until 1960. The headquarters of ICL, with the same name, now occupies the site (Fanning, *op. cit.*, 15–16).

66 XXIV, 1.

67 *Chr*, 151.

68 CH1, ii, 33.

69 CH1, ii, 34.

70 The mirror of 4-ft focal length and the complete 7-ft reflector that he supplied to J. H. Schroeter for his private observatory at Lilienthal are exceptional in being put to serious use. See Maurer, nos C-7 and C-8.

71 Papers in private possession.

72 RAS MS W.1/1, 202–5. See Maurer 15 for a complete list as of 1794.

73 XI, 28, from *Journal of Mrs. Papendiek*, whose father held the lease of the house at Slough before it was taken over by William.

74 William to J. E. Bode, 20 July 1785, RAS MS W.1/1, 134–5; William Watson, Jr, to William, 5 June 1786; [Barnaba Oriani], *Un Viaggio in Europa nel 1786: Diario di Barnaba Oriani Astronomo Milanese* (Florence, 1994), 138 ("Suo fratello lavora i telescopi de vendere quando non v'e musica a Bath"). See John Tracy Spaight, "Alexander Herschel as Telescope Maker", *JHA*, xxxiv (2003), 95–96.

75 XI, 20–21.

76 William, writing some time after the event,

says: "... my present situation being much more limited with regard to income than my former one at Bath, I thought it prudent to request the favour of the President of the R.S. to make an application to the King. His Majesty most graciously granted my petition..." (RAS MS W.5/12.1, 69). According to this account, it was then that William invited the King to choose between a 30-ft and a 40-ft reflector. But the letter of application is quite specific that it is a 40-ft that is proposed, and the list of expenses is based on this size. William may therefore have forgotten the true sequence of events. Alternatively, the application does envisage mirrors of either 3-ft or 4-ft diameter, and it may have been that the King was asked to choose between these two alternatives. Like all grant-giving bodies, he was unlikely to opt for the smaller if he could have the larger for his money. Since the metal for the mirrors was a major item of expense, it is indeed curious that William ventured to quote a price before the size of the mirrors had been decided, but he had originally planned 4-ft mirrors to be cast in his basement in his Bath days and it is unlikely that his current ambitions would have been satisfied with anything less.

77 RAS MS W.5/12.1, 69.

78 Bennett, 93–95. The fine water-colour of the completed instrument, sent with the assembly instructions, is reproduced in Michael Hoskin (ed.), *The Cambridge Illustrated History of Astronomy* (Cambridge, 1997), 247.

79 In 1840 William Parsons, future third Earl of Rosse, succeeding in building a reflector whose mounting was based on those of Herschel, with 3-ft mirrors and 26-ft focal length. See Michael Hoskin, "The Leviathan of Parsonstown: Ambitions and Achievements", *JHA*, xxxiii (2002), 57–70.

80 RAS MS W.1/5.1.

81 George, Letter 181.

82 In the list of expenses erroneously printed with George, Letter 432. It belongs in fact to Letter 369.

83 XI, 25.

84 William to Charles Blagden, 27 November 1786, RAS MS W.1/1, 151–3.

85 *Chr*, 145.

86 William Watson, Jr, to William, 11 November 1785, RAS MS W.1/13.W.40.

87 XI, 4. Although Bath was an easy journey from Slough, William and Caroline were not present at the wedding (and Caroline's words suggest they were not invited).

88 XII, 11.

89 XII, 11.

90 XII, 10–11.

91 XII, 13.

92 XII, 13.

93 XII, 16.

94 XII, 15–16.

95 RAS MS C.1/1.1, 38.

96 *Chr*, 169.

97 XI, 19b; *Mem*, 309 (Caroline to John's wife, 10 January 1840).

98 George, Letter 379.

99 As mentioned above, the details are mistakenly attached to George, Letter 432, in the printed edition.

100 RAS MS W.1/5.2(i).

101 Sir Joseph Banks to William, 23 August 1787, RAS MS W.1/13.B.20.

102 Caroline to John, April 1827, *Mem*, 209.

103 William Watson, Jr, to William, 7 September 1787, RAS MS W.1/13.W.50.

104 Caroline to John, April 1827, *Mem*, 211.

105 See Roy Porter, *Mind Forg'd Manacles: Madness and Psychiatry in England from Restoration to Regency* (London, 1987).

106 He was made a knight of the Royal Guelphic Order, and the oil painting by William Artaud shows him wearing the insignia. The Order was instituted by the Prince Regent on 12 August 1815, a few months after Hanover had become a formal kingdom following the defeat of Napoleon. It was awarded for distinguished services to the United Kingdom or Hanover, and William, as a Hanoverian resident in England, was a very suitable appointee. His son John became a knight in 1831. As a British Order, it lapsed when the crowns were separated in 1837, following the accession of a female to the British

throne.

107 In a letter dated 21 September 1820, W.7/10. The document confirming Caroline's pension was sold at Sotheby's on 3/4 March 1958 as part of Lot 483.

108 George, Letter 570.

109 George, Letter 432.

110 CH1, ii, 49–50.

111 XIII, 1.

112 The building was enlarged during the nineteenth century and became a girls' school; it has since been pulled down (information courtesy of Cdr Anthony Fanning).

113 *Chr*, 173.

114 XIII, 7.

115 The Baldwins had long been owners of the Crown Inn. In 1701 William Baldwin was granted a long lease on the property, and he was succeeded in 1703 by his son Thomas, who instituted the first daily coach between London and Bath. The gardens of the Inn extended as far as the Herschel home. Thomas's son Adee left the Crown Inn to his widow Elizabeth, on whose death it was to pass to their daughter Mary (who married William). At some stage the lease was converted to ownership, for William's property at his death included both the Crown Inn and the adjacent house, and the Crown Inn was still owned by William's son John at his death. The distance from the Herschel home to the inn was some 200 yards. See *Slough*, 55–56.

116 J. T. Spaight has located Pitt's will, and has kindly supplied the writer with a copy.

117 *Chr*, 174. Clerke, 44, says that Mary was 38 when she married William on 8 May 1788.

118 Paul died in February 1793, *Chr*, 238. Mary lived on until the beginning of January 1832, M. J. Crowe (ed.), *A Calendar of the Correspondence of Sir John Herschel* (Cambridge, 1998), 2505.

119 Mrs Baldwin in fact died on 22 October 1798, XXI, 9.

120 *Chr*, 177.

121 *Chr*, 174.

122 *Chr*, 174.

123 William Watson, Jr, to William, 24 March 1788, RAS MS W.1/13.W.51; *Chr*, 175.

124 *Chr*, 173.

125 *Chr*, 176.

126 Caroline to Mary, 14 October 1824, *Mem*, 178.

127 CH1, ii, 50.

128 CH1, ii, 49.

129 *Chr*, 233.

130 Anna disappears from the record. She was married in 1732 and so must have been in her mid-seventies, or older, at the time of William's wedding.

131 On pp. 133–4 of the shortlived *Berlinische Musikalische Zeitung* for 1793, there is an editorial item about music in Hanover, based on a letter received from there. It states: "Der Vice-Konzertmeister, Jacob Herschel, ward im vorigen Jahre im Felde erwürgt gefunden, und schon lange ist keine Stelle wieder besetzt worden ..." ("The Vice-Concertmaster Jacob Herschel was found last year strangled in the field, and the position has since been vacant ..."). Arndt Latusseck, to whom I am indebted for this information, remarks that "im Felde sein" almost always implies "in the field [of battle]", and indeed the Hanoverian writer complains of the lack of oboists due to the present "Kriegsumstände" ("circumstances of war"). It is hard to imagine Jacob voluntarily taking part in military service, and it seems more likely that he was a civilian who found himself in the wrong place at the wrong time.

132 Money advanced and salary paid by the King for the great telescope between February 1786 and July 1790 amounted to £2947 10s., according to George, Letter 456fn. This included payment for Caroline.

133 Caroline to John's wife, 14 October 1824, *Mem*, 178.

Chapter 4

1 XXI, 2.

2 RAS MS W.5/12.1, 95, 8 December 1788.

3 *Chr*, 294.

4 *Chr*, 178.

5 Accounts in private possession.

6 All these amounts are cited in extracts from her accounts, in private possession.

7 Caroline to Mary, 14 October 1824, *Mem*, 178.

8 *Chr*, 146, from *Journal of Mrs. Papendiek*.

9 Agnes M. Clerke, entry in *Dictionary of National Biography*.

10 *Chr*, 308–9.

11 *Chr*, 310.

12 *Chr*, 311.

13 WH, "Observations Tending to Investigate the Nature of the Sun", *PT*, xci (1801), 265–318, p. 265.

14 WH, "Investigation of the Powers of the Prismatic Colours to Heat and Illuminate Objects", *PT*, xc (1800), 255–83, p. 272.

15 RAS MS W.1/13.B.34, dated 9 April 1800.

16 Bennett, 97.

17 RAS MS W.1/4; W.1/1, 290.

18 RAS MS W.1/13.B.178.

19 *Chr*, 143. To Burney, Caroline was William's "ingenious and accomplished Sister", RAS MS W.1/13.B.176.

20 RAS MS W.1/13.B.179.

21 It is not to be found in Walmer Castle, nor do the fragmentary records there give any clue as to its fate (Rowena Shepherd, personal communication, June 2002).

22 XX, 12.

23 William to Caroline, 7 November 1800, *Mem*, 106.

24 The only numerical errors that have ever been identified in her work occur in the catalogue of 2500 nebulae she prepared for John when she was in her seventies and which is now in the Royal Society Library (see Chapter 5). Slips in a single digit occur, for example, in nebulae III 325 and 381.

25 John Flamsteed, *Historia Coelestis Britannica* (London, 1725).

26 WH, "A Third Catalogue of the Comparative Brightness of the Stars", *PT*, lxxxvii (1797), 293–324, p. 293; *Chr*, 256.

27 *Ibid.*

28 Caroline to Nevil Maskelyne, September 1798, *Chr*, 257; Caroline to Francis Baily, 2 April 1835, *Mem*, 273.

29 From p. 3 of his introduction to the first section. The second section, which identified where in vol. ii any given star of the Catalogue was to be found, ran to 76 folio pages of symbols, a typesetter's and proof-reader's nightmare. The work was published in London in 1798.

30 Caroline to Nevil Maskelyne, September 1798, *Chr*, 257.

31 Her calculations survive as RAS MS C.2/4.5, and the catalogue of omitted stars arranged in order of Right Ascension for the use of Maskelyne as C.2/5.1 and C.2/5.2.

32 *Mem*, 100.

33 *Mem*, 109. She received the Bode catalogue on 24 September 1801, when she had progressed as far as Sweep 387. "I was then by this acquisition brought into some confusion and if there had been time for calculating the 387 Sw over again the beginning would not have been so crowded &c.", memorandum dated 7 March 1834, RAS MS C. 4/3, f. 18.

34 RAS MS C.3/2.3.

35 Nevil Maskelyne to William, 10 January 1789, RAS MS W.1/13.M.37.

36 Nevil Maskelyne to William, 16 January 1790, RAS MS W.1/13.M.44. He continues: "... after your example and under your auspices."

37 Francis Wollaston to William, summer 1789, RAS MS W.1/13.W.194.

38 Karl Seyffer to Caroline, 10 May [?1793], *Mem*, 92.

39 M.-A. Pictet to William, 20 April 1799, RAS MS W.1/13.P.24.

40 Jerôme de Lalande to Caroline, 12 July 1790, *Chr*, 251.

41 To take one example among many, in a letter to William of 1790 Jerôme de Lalande sends "Mille respects à la Savante Miss", RAS MS W.1/13.L.18.

42 Jerôme de Lalande to William, 9 September 1788, RAS MS W.1/13.L.12: "Mille tendres respects à la savante miss, dont je parle

souvent avec entousiasme."

43 For details of this sweeper, see above. Caroline retained an affection for the small sweeper, which she could use while seated. The large sweeper, whose focal length exceeded her own height, she used while standing on steps, XVIII, 16.

44 RAS MS C.1/1.4.

45 RAS MS C.1/1.4.

46 On 13 October 1793, RAS MS C.1/1.2.

47 Jérôme de Lalande to William, 26 April 1787, RAS MS W.1/13.L.5.

48 Jérôme de Lalande to William, 1 April 1792, RAS MS W.1/13.L.23. The prizes were "pour l'ouvrage le plus utile ou la decouverte la plus important pour les sciences ou les arts".

49 Jérôme de Lalande to William, 16 August 1788, RAS MS W.1/13.L.11: "Je n'oublierai jamais surtout la nuit du 5 Aout, j'ai dit à tout le monde que jamais je n'en avois passé d'aussi agréables, sans excepter même celles d'amour."

50 RAS MS C.1/1.4.

51 David Hughes, "Caroline Lucretia Herschel — Comet Huntress", *Journal of the British Astronomical Association*, cix/2 (1999), 78–85, p. 82.

52 RAS MS C.1/1.2 91.

53 Nevil Maskelyne to Caroline, 27 December 1788, *Chr*, 246.

54 Caroline to Sir Joseph Banks, 17 August 1797, *Mem*, 94–95.

55 Simon Schaffer, "'The Great Laboratories of the Universe': William Herschel on Matter Theory and Planetary Life", *JHA*, xi (1980), 81–111, pp. 96–100.

56 Caroline to John, 3 May 1825, *Mem*, 190.

57 *Mem*, 73.

58 *Chr*, 158.

59 *Mem*, 73.

60 Bennett, 89.

61 RAS MS W.5/12.1, 93; W.5/14.1, f. 19.

62 RAS MS W.5/14.1, f. 35v.

63 RAS MS W.5/12.3, expt 3.

64 Quoted in *Chr*, 168.

65 On the Windsor sheet of the first edition, published in 1830.

66 XII, 35.

67 WH, "On Nebulous Stars, Properly So Called", *PT*, lxxxi (1791), 71–88.

68 George Griesbach (1757–1824) joined the band in May 1778, his brother Henry (1762–1832) in 1780, and Frederick (1769–1825) and William (1772–) followed suit in the next few years (VI, appendix by George Griesbach; Highfill *et al.*, *A Biographical Dictionary of Actors*). They must have added to the gaiety of nations as they went about in scarlet coat, waistcoat and breeches, cocked hat, and sword. Charles joined the band no later than 1794 (*ibid.*), although the appendix summarizes George as saying he joined after their mother's death which took place in April 1803 (*Chr*, 316).

69 Jérôme de Lalande to William, 30 May 1785, RAS MS W.1/13.L.3.

70 Jean-Dominique de Cassini to William, 15 February 1785, RAS MS W.1/13.C.6.

71 Jérôme de Lalande to William, 8 November 1786, RAS MS W.1/13.L.4.

72 RAS MS W.1/13.W.2/4.

73 RAS MS W.2/1.1, f. 1.

74 RAS MS W.2/4, f. 1v.

75 RAS MS W.2/4, f. 2v.

76 William to Sir Joseph Banks, 21 October 1789, RAS MS W.1/1, 181.

77 WH, "Account of the Discovery of a Sixth and Seventh Satellite of the Planet Saturn", *PT*, lxxx (1790), 1–20, p. 1.

78 In his edition of William's papers, Dreyer (i, 373, 375) is careful to clarify the position in footnotes.

79 WH, "On the Power of Penetrating into Space by Telescopes", *PT*, xc (1800), 49–85, p. 85.

80 WH, "Astronomical Observations Relating to the Sidereal Part of the Heavens", *PT*, civ (1814), 248–84, p. 275.

81 WH, "A Series of Observations of the Satellites of the Georgian Planet", *PT*, xv (1815), 293–362, p. 295–6.

82 XXV, 7–9.

83 William wrote in October 1810, "A mirror of this size cannot last above 3 or 4 years without being repolished", RAS MS W.2/2.8, f. 3.

84 *Mem*, 113.

85 There is no mention in the Polishing Record after 1809 (Dreyer, i, p. liv: Bennett,

ref. 132), and Dreyer wonders if in subsequent years preparations were made but the work then abandoned.

86 *Mem*, 123. XXIV, 32 confirms this polishing was the last. *Chr*, 342 says that the mirror was in the polishing room in the summer of 1813, and that in addition regular polishing was continued "up to June or perhaps September, 1814".

87 *Mem*, 124.

88 XXIV, 33.

89 XXV, 8.

90 Dreyer, i, p. lv.

91 Caroline to John, April 1827, *Mem*, 209–10.

92 Caroline to John's wife, 6 September 1833, *Mem*, 259.

93 *Mem*, 95.

94 Caroline to John's wife, 6 September 1833, *Mem*, 259.

95 *Mem*, 98.

96 *Mem*, 98–99.

97 *Mem*, 100. According to *Slough*, 131, 138, Caroline at one time lodged at Elizabeth Cottage, now 208 High Street, and this may have been the tailor's premises.

98 *Mem*, 104.

99 *Mem*, 105.

100 *Mem*, 107.

101 *Mem*, 108.

102 *Mem*, 109.

103 XXI, 8.

104 *Mem*, 101.

105 William first took a house in Bath on 11 February 1799, XXI, 10.

106 On 4 Sept. 1799 (when he observed the nebula M 2), and 28 Dec. of the same year (M 74). See Michael Hoskin, "Herschel's 40-ft Reflector: Funding and Functions", *JHA*, xxxiv (2003), 1–32. In his defence William might have said that he had made for himself a 7-ft reflector in "skeleton form", "the tube and stand consisting of bars to be screwed together and taken asunder so as to lie in a small compass for convenient carriage" (Dreyer, i, p. lv), and this he sometimes took with him on his travels. But it is hard to see the skeleton 7-ft as an instrument for serious research, and he used it more to entertain the friends he visited.

107 *Mem*, 104.

108 In November 1799 Watson wrote to William in some alarm, having heard that William intended to part with his house there. But Alexander had reassured Watson that "you & Mrs Herschel proposed still to visit Bath but meant to have a house in the town". Watson tells William of a three-windowed house available for £55 rent plus £20 taxes. William Watson, Jr, to William, 2 March 1803, RAS MS W.1/13.W.64.

109 RAS MS W.1/13.W.77, dated 2 March 1803.

110 *Mem*, 105.

111 *Mem*, 105.

112 *Mem*, 106.

113 *Mem*, 106.

114 *Mem*, 107.

115 *Mem*, 112.

116 *Mem*, 115.

117 *Mem*, 115.

118 *Mem*, 122–3.

119 Papers in private possession.

120 XXIII, 6.

121 XXIII, 7.

122 XXIII, 6–7.

123 *Mem*, 22.

124 VI, 3.

125 Dietrich to William, 2 March 1806, *Chr*, 318.

126 See ref. 7.

127 *Mem*, 115.

128 *Mem*, 116.

129 Memorandum now in the Harry Ransom Library, Humanities Research Center, University of Texas at Austin.

130 *Mem*, 119.

131 *Mem*, 125.

132 *Mem*, 136.

133 *Mem*, 136.

134 XXIV, 43.

135 XXIV, 47.

136 *Mem*, 125.

137 XXV, 1.

138 VI, 3; written in Feb 1822 (marginal note).

139 Title of the first chapter of Günther Buttmann, *The Shadow of the Telescope: A Biography of John Herschel*, English transl. (London, 1974).

140 *Chr*, 329. For details of the inn, see above. The house lay between William's home

and the Crown Inn.

141 *Slough*, 56.

142 *Chr*, 330.

143 *Chr*, 330.

144 XXIV, 1.

145 XXIV, 2.

146 *Mem*, 120.

147 Accounts, entries for 29 July 1806 and 6 April 1808 (papers in private possession).

148 XXIV, 26.

149 If Caroline had been living in Mrs Baldwin's old house, as the author of *Chr* thinks likely, the enforced letting of an independent property in the ownership of William would make no sense.

150 XXIV, 30.

151 WH, "Astronomical Observations Relating to the Construction of the Heavens", *PT*, ci (1811), 269–336, Section 19.

152 *Ibid.*, Introduction.

153 *Mem*, 112.

154 XXV, 11–12; *Mem*, 129–30.

155 XXV, 24.

156 XXV, 21.

157 They were needed to complete a set of 72 published papers by William. These were bound in five volumes, and eventually auctioned by Sotheby's on 4 March 1958 as Lot 393; the set fetched a derisory £45.

158 Buttman, *Shadow*, 17.

159 RAS MS W.5/12.4, entry for 1 April 1816 (in John's hand).

160 Buttman, *Shadow*, 20.

161 Dreyer, i, p. lv.

162 XXV, 18.

163 XXV, 18.

164 "It was constructed in the year 1820, under the joint superintendence of my father and myself, on the model of that used by him in his 'Sweeps of the Heavens,' whose place it was intended to supply, the wood-work of the latter being greatly decayed by age. It was therefore dismounted, and the less perishable part of its materials employed in the construction of its successor", J. W. F Herschel, "Account of Some Observations Made with a 20-feet Reflecting Telescope", *Astronomical Society Memoirs*, ii (1826), 459–97.

165 XXV, 19. On another occasion (p. 459 of his great volume *Results of Observations ...* (London, 1847) reporting the results of his work at the Cape of Good Hope), John saw the new 20-ft as constructed "under the joint supervision of my father and myself".

166 RAS MS J.1/1, entries for 29 and 30 May 1821.

167 RAS MS J.1/1.

168 *Mem*, 132–3.

169 Originals now in the Harry Ransom Library, Humanities Research Center, University of Texas at Austin. The first instalment, covering her life in Hanover, she completed on 16 March 1822. She resumed work when she was in Hanover, and by 1827, when Dietrich died, she had written of the Bath years and the early months in Datchet. She then wrote for John an account of the middle 1780s, the high point of her partnership with William, but the autobiography ends on the day of William's marriage.

170 XXV, 23–24.

171 *Mem*, 137.

Chapter 5

1 Caroline to John's wife, 5 April 1840, *Mem*, 312.

2 *Mem*, 138.

3 The term John uses in a letter to Francis Baily dated 12 March 1828, Royal Society papers HS 3.74.

4 Caroline to John's wife, 9 May 1828, *Mem*, 221.

5 Herschel/M1090, 1, Harry Ransom Library, Humanities Research Center, University of Texas at Austin.

6 Caroline to John, 16 August 1827, *Mem*, 216.

7 Buttman, *Shadow*, 29–33.

8 Caroline to John, 8 August 1826, *Mem*, 200.

9 *Mem*, 136. In her letter to John of 18 April 1832, British Library Egerton 3761, ff. 154–7, she is more specific: "For in consequence of a letter I received shortly before I left England in which my Brother [Dietrich] acquainted me of the deranged affairs of his son in law Dr Richter ... I made him by return of the post the offer of the whole of my property (which perhaps you know) was 500£ in the 3 p[er]c[en]t." I thank Emily Winterburn for drawing my attention to this letter, which is overlooked by Crowe.

10 She did not, it seems, know of the annuity she was to receive under William's will since she later describes this as "unexpected", *Mem*, 139.

11 Herschel/M1090, 1, Harry Ransom Library, Humanities Research Center, University of Texas at Austin.

12 John to Sir David Smith, 25 September 1822, Royal Society Library, HS 19.37.

13 John to Dietrich, 6 September 1822, Royal Society Library, HS 19.30.

14 *Mem*, 138.

15 *Mem*, 139.

16 John to Charles Babbage, 16 October 1822, Royal Society Library, HS 2.179.

17 *Mem*, 139.

18 Caroline to Mary, 21 October 1822, *Mem*, 152.

19 Caroline to John, 25 September 1827, *Mem*, 219.

20 *Ibid.*, *Mem*, 218.

21 Caroline to Mary, 30 October 1822, *Mem*, 154–5.

22 Caroline to Mary, 10 July 1827, *Mem*, 215.

23 Caroline to Mary, 12 November 1822, *Mem*, 156.

24 Caroline to Mary, 30 October 1822, *Mem*, 155.

25 Caroline to Mary, 12 November 1822, *Mem*, 157.

26 XXVI, 20 gives the date as in the 1840s, and evidently Constance Lubbock found it impossible to be more precise before *Chr* went to press. Mme Beckedorff was still alive when Caroline penned page 26 of her 1840s autobiography.

27 *Chr*, 376.

28 XXVI, 13.

29 Caroline to John, 27 February 1823, *Mem*, 163.

30 The outline details of William's will were published in *Gentleman's Magazine*, xcii (1822), 650, and a full account is given by P. D. Hingley, "The Will of Sir William Herschel", *Astronomy & Geophysics*, xxxix (1998), 3.7.

   In the original will, dated 1818, he left £20,000 to John, together with his properties including the Crown Inn, but under an 1821 codicil John received an extra £5,000. He left £100 annually to Caroline, £2,000 to Dietrich, £20 each to his nephews and nieces, his astronomical materials to John "for the purpose of enabling him to pursue the Study of Astronomy which by reason of my Age I have for some time past declined", and the residue to his widow. The total valuation was declared to be less than £60,000.

31 Caroline to John, postscript to Mary, 1 November 1824, *Mem*, 180.

32 John to Peter Stewart, 22 October 1834, Royal Society Library HS 25.4.2.

33 John to Caroline, 1 August 1823, *Mem*, 169.

34 *Chr*, 367 says that Caroline took with her the Books of Sweeps and the "Catalogue of 2500 Nebulae", but this title must be an error for her catalogue of the stars featuring in sweeps.

35 Caroline to John, 11 August 1823, *Mem*, 171. In 1816 Caroline had noted, "I took the opportunity of working on my MSs Catalogue at those times I was left without employment", XXIV, 42, and Lubbock thinks she is referring to the zone catalogue. But Caroline must be referring instead to a catalogue of manuscripts in William's library. *Cf.* her entry for 5 February 1817, where she "had restored order in his Library and workrooms", XXV, 2.

36 The catalogue is MS 279 in the Royal Society Library, and there is another copy of Caroline's explanatory preface in RAS MS C.4/4.

37 Caroline to John, 25 September 1824, *Mem*, 176.

38 John to John Grahame, 25 October 1824, Royal Society Library HS 8.320.

39 Caroline to John, 14 January 1825, *Mem*, 181.

40 Caroline to John, 27 March 1825, *Mem*, 186. See RAS MS C.3/2.3. The title, adapted from the "Catalogue of the stars which have been observed by Wm Herschel in a series of sweeps" on which it was based, was: "A catalogue of the nebulae which have been observed by William Herschel in a series of sweeps; brought into zones of N[orth] P[olar] Distance and order of R[ight] A[scension] for the year 1800, by applying to the determining stars the variations given in Wollaston's or Bode's catalogues". For a description see Dreyer, i, pp. lxiii–lxiv.

41 MS 279.

42 A sheet is loosely inserted towards the back of the catalogue.

43 Cited by Clerke, 132.

44 Cited by Clerke, 132.

45 John to Caroline, 18 April 1825, *Mem*, 188.

46 John to Caroline, 4–11 May 1827, *Mem*, 213.

47 John to Caroline, 18 April 1825, *Mem*, 188.

48 Caroline to John, 1 February 1826, *Mem*, 196, 198.

49 Caroline to John, 20 September 1825, *Mem*, 191–2.

50 Caroline to John, 8 August 1826, *Mem*, 200.

51 Caroline to John, 24 December 1826, *Mem*, 208.

52 At 376 Braunschweigerstrasse, *Mem*, 214; Agnes Clerke, entry in *Dictionary of National Biography*.

53 Caroline to John, 8 August 1826, *Mem*, 200.

54 Caroline to John, 26 December 1822, *Mem*, 161.

55 RAS MS C.1/1.3, 111.

56 Dreyer, i, p. lv.

57 Caroline to John, 11 August 1823, *Mem*, 171.

58 Dreyer, i, p. lv.

59 Caroline to John's wife, 3 August 1840, *Mem*, 313.

60 On the foundation of the Royal Astronomical Society, see J. L. E. Dreyer *et al.*, *A History of the Royal Astronomical Society, 1820–1920* (London, 1923).

61 As shown by his comment quoted in Caroline's record, RAS MS J.1/1.

62 *Mem*, 225. The gold medal is now in Girton College, Cambridge.

63 Caroline to John, 3 June 1828, *Mem*, 228.

64 Caroline to John, 21 August 1828, *Mem*, 232.

65 John to Caroline, 28 May 1828, *Mem*, 227.

66 Caroline to John, 23 June 1828, *Mem*, 230–1.

67 Caroline to Mary, 16 November 1829, *Mem*, 238.

68 Caroline to John, 18 April 1832, British Library Egerton 3761, ff. 154–7.

69 *Ibid.*

70 *Ibid.*

71 *Ibid.*

72 Caroline to John, 18 June 1830, *Mem*, 240–1.

73 RAS MS C.2/8.

74 RAS MS C.2/9.

75 RAS MS C.2/8, on the reverse of the engraved portrait.

76 John to his wife, 19 June 1832, *Mem*, 254–5.

77 Caroline to John's wife, 4 December 1832, *Mem*, 256.

78 Caroline to Mary, postscript to John, 1 August 1833, *Mem*, 258.

79 John to Caroline, 6 June 1834, *Mem*, 266.

80 Caroline to John, 11 September 1834, *Mem*, 269.

81 John to Caroline, 22 February 1835, *Mem*, 270.

82 Dietrich's daughter, Anna Elise Kipping, to John, 18 January 1848, *Mem*, 346.

83 Cited by Richard A. Proctor in "Sir John Herschel as a Theorist in Astronomy", reprinted from *The St Paul's Magazine* for June 1871 in his *Essays in Astronomy* (London, 1872), 8–28, pp. 16–17.

84 Occasional papers marked for destruction survive among the William Herschel papers in the archives of the Royal Astronomical Society.

85 Michael Hoskin, "John Herschel's Cosmology", *JHA*, xviii (1987), 1–34, pp. 3–5.

86 John to Caroline, 7 March 1841, Harry Random Library, Humanities Research

Center, University of Texas at Austin.

87 *Report of the Fifteenth Meeting of the British Association for the Advancement of Science* (London, 1846), p. xxxvi.

88 Caroline to John, 4 August 1842, *Mem*, 327. Caroline speaks of "Lord Queenstown". The future Earl of Rosse, of Parsonstown in King's County, was known by the courtesy title of Lord Oxmantown.

89 Buttman, *Shadow*, 119.

90 John to W. H. Sykes, 12 April 1839, Royal Society Library, HS 17.140.

91 William James Herschel was born in January 1833.

92 *Mem*, 294.

93 *Mem*, 295.

94 Caroline to John's wife, 3 March 1842, *Mem*, 323–4.

95 Caroline to John, 8 August 1826, *Mem*, 200.

96 Caroline to Mary, 3 March 1829, *Mem*, 235.

97 *Briefwechsel zwischen A. v. Humboldt und C. F. Gauss*, ed. by K. R. Biermann (Berlin, 1977), 59.

98 *Mem*, 304. The head of the page gives an erroneous date (1839) and this has misled some writers; but a few days before, Caroline had been visited by Alexander von Humboldt, and the year of this visit (1837) is known from his writings. See Margaret Bullard, "My Small Newtonian Sweeper — Where Is It Now?", *Notes and Records of the Royal Society of London*, xvii (1988), 139–48, p. 144.

99 Thomas Galloway writes to John on 25 April 1840 notifying him of the books found in the crate and asking whether they were intended for the Society or should be returned to him, Royal Society Library, HS 8.9.

100 John to Francis Baily, 14 April 1840, Royal Society Library, HS 25.8.26.

101 Francis Baily to John, 20 April 1840, Royal Society Library, HS 3.181.

102 Thomas Galloway to John, 25 April 1840, Royal Society Library, HS 8.9. The gift was reported to the meeting of 8 May, *Monthly Notices of the Royal Astronomical Society*, v (1843), 79. As noted above, the instrument is now in the Science Museum in London.

103 Caroline to John's wife, 5 April 1840, *Mem*, 312. On 3 August Caroline told John's wife, "... if my nephew thought the seven-foot telescope worth the acceptance of the Royal Astronomical Society, it is well!", *Mem*, 313.

104 *Mem*, 305.

105 Ulrich Friedrich Hausmann (1776–1847) has been confused with the more famous Johann Friedrich Ludwig Hausmann (1782–1859), who however lived in Göttingen. Ulrich Friedrich was a resident of Hanover in later life, but he may possibly have met Caroline when he was in England as a war refugee in the early years of the century. I thank Arndt Latusseck for this information.

106 Bullard, *op. cit.* (ref. 98), 141.

107 Memorandum in the Harry Ransom Library, Humanities Research Center, University of Austin at Texas.

108 *Ibid.*

109 Caroline to John, 1 September 1840, *Mem*, 316.

110 John to Caroline, 5 August 1845, Harry Ransom Library, Humanities Research Center, University of Texas at Austin.

111 *Chr*, 373.

112 *Mem*, 336.

113 G. B. Airy to John, 13 October 1846, Royal Society Library, HS 1.168.

114 *Chr*, 376.

115 John to Caroline, 11 July 1847, *Mem*, 342.

116 Dietrich's daughter, Anna Elise Kipping, to John, 18 January 1848, *Mem*, 346.

117 John to Augustus DeMorgan, 3 February 1848, Royal Society Library, HS 23.10.

118 *Mem*, 347.

Chapter 6

1 *Chr*, 369.

2 *Mem*, 223–4.

3 *Mem*, 224–5.

4 Fanny Burney (Madame d'Arblay), *Diary*, edn of 1842, iii, 442.

5 *Mem*, 136.

6 VI, 3.

7 VI, 3, written in February 1822. No doubt Caroline was already worrying about what to do when William died.

8 *Mem*, 136.

9 VI, 27.

10 *Mem*, 167.

11 Clerke, 139.

12 Clerke, 139.

13 *Chr*, 57.

14 John to Rudolph Wolf, 25 March 1866, Royal Society Library HS 19.299; *Chr*, 58.

15 *Chr*, 369.

Indexes

(1) The Herschel Family

(2) General